International Conference on Information Technology and Digital Applications (ICITDA 2017)

IOP Conference Series: Materials Science and Engineering
Volume 325

Yogyakarta, Indonesia
8 - 9 November 2017

Editors:

Taufiq Hidayat
Ahmad Fathan Hidayatullah

ISBN: 978-1-5108-5984-5
ISSN: 1757-8981

Printed from e-media with permission by:

Curran Associates, Inc.
57 Morehouse Lane
Red Hook, NY 12571

Some format issues inherent in the e-media version may also appear in this print version.

Printed by Curran Associates, Inc. (2018)

For permission requests, please contact the Institute of Physics at the address below.

Institute of Physics
Dirac House, Temple Back
Bristol BS1 6BE UK

Phone: 44 1 17 929 7481
Fax: 44 1 17 920 0979

techtracking@iop.org

Additional copies of this publication are available from:

Curran Associates, Inc.
57 Morehouse Lane
Red Hook, NY 12571 USA
Phone: 845-758-0400
Fax: 845-758-2633
Email: curran@proceedings.com
Web: www.proceedings.com

Table of contents

Volume 325

International Conference on Information Technology and Digital Applications (ICITDA 2017)

8–9 November 2017, Yogyakarta, Indonesia

Accepted papers received: 21 February 2018
Published online: 13 March 2018

Preface

Papers

IOP Conf. Series: Materials Science and Engineering **325** (2018) 011001 doi:10.1088/1757-899X/325/1/011001

PREFACE

Welcome to International Conference on Information Technology and Digital Applications (ICITDA) 2017. This is an annual international forum for the exchange of ideas, knowledge, and experience on the latest development in the field of information technology and digital applications among researchers from academia, engineers, and practitioners from industries and also expected to enable opportunities for collaboration among the participants to advance the theory and practice in the field of information technology.

ICITDA 2017 is the second ICITDA in collaboration with Rajamangala University of Technology Thanyaburi, Thailand, and Youngsan University, South Korea. This conference also the third international conference organized by the Department of Informatics, Faculty of Industrial Technology, Universitas Islam Indonesia.

This year, 43 papers were reviewed by academia and practitioner from Indonesia, United States, Australia, Thailand, and United Kingdom, and 29 papers from Indonesia, Thailand, United States, Philippines and also South Korea are eligible to be presented in ICITDA 2017, 8-9 November 2017 at Harper Mangkubumi Hotel, Yogyakarta, Indonesia.

We hope, with your participation in ICITDA 2017, we can grow positive impact of information technology to the community and also strengthen the researcher network in Southeast Asia.

Ari Sujarwo, Chair

November 2017

IOP Conf. Series: Materials Science and Engineering **325** (2018) 011001 doi:10.1088/1757-899X/325/1/011001

CHAIRS

Host

- Department of Informatics, Faculty of Industrial Technology, Universitas Islam Indonesia

Editors

- Taufiq Hidayat
- Ahmad Fathan Hidayatullah

Organizing Committee

- Ari Sujarwo (Chairman of The Committee)
- Hendrik
- Lizda Iswari
- Ahmad M. Raf'ie Pratama
- Ridho Rahmadi
- Taufiq Hidayat
- Dhomas Hatta Fudholi
- Ahmad Fathan Hidayatullah
- Almed Hamzah
- Moh. Idris
- Fayruz Rahma
- Zulfahmi Kesuma A.

Program Committee

- Rinaldi Munir (Institut Teknologi Bandung, Indonesia)
- Rila Mandala (Institut Teknologi Bandung, Indonesia)
- Burasakorn Yoosooka (Rajamangala University of Technology Thanyaburi, Thailand)
- Reyer Zwiggelaar (Aberystwyth University, United Kingdom)
- Ermatita (Universitas Sriwijaya, Indonesia)
- Husni S. Sastramihardja (Universitas Esa Unggul, Indonesia)
- Jefferson Tan (Monash University, Australia)
- Sritrusta Sukaridhoto (Politeknik Elektronika Negeri Surabaya, Indonesia)
- Agfianto Eko Putra (Universitas Gadjah Mada, Indonesia)
- Nur Aini Rakhmawati (Institut Teknologi Sepuluh Nopember, Indonesia)

IOP Conf. Series: Materials Science and Engineering **325** (2018) 011001 doi:10.1088/1757-899X/325/1/011001

- Wayan Suparta (Universiti Kebangsaan Malaysia, Malaysia)
- Fitra A. Bachtiar (Universitas Brawijaya, Indonesia)
- Taufiq Hidayat (Universitas Islam Indonesia, Indonesia)
- Dhomas Hatta Fudholi (Universitas Islam Indonesia, Indonesia)
- M. Andri Setiawan (Universitas Islam Indonesia, Indonesia)
- Izzati Muhimmah (Universitas Islam Indonesia, Indonesia)
- Irving Vitra Paputungan (Universitas Islam Indonesia, Indonesia)
- Syarif Hidayat (Universitas Islam Indonesia, Indonesia)
- R. Teduh Dirgahayu (Universitas Islam Indonesia, Indonesia)
- Ahmad Luthfi (Universitas Islam Indonesia, Indonesia)
- Yudi Prayudi (Universitas Islam Indonesia, Indonesia)
- Ridho Rahmadi (Universitas Islam Indonesia, Indonesia)
- Ahmad M. Raf'ie Pratama (Universitas Islam Indonesia, Indonesia)
- Fathul Wahid (Universitas Islam Indonesia, Indonesia)
- Hendrik (Universitas Islam Indonesia, Indonesia)
- Kurniawan Dwi Irianto (Universitas Islam Indonesia, Indonesia)

IOP Conf. Series: Materials Science and Engineering **325** (2018) 011001 doi:10.1088/1757-899X/325/1/011001

Keynote Speeches

Dr. Ahmed Imran

University of New South Wales, Australia

Smart Systems and ICT4D: prospects, fads and traps

With the proliferation and rapid development of Information Communication Technology (ICT) and smart systems, many of the development agendas in developing countries had been exploring ICT based solutions to tackle the wicked problems of those countries. While ICT emerged as the potential enabler and tool to solve many problems including poverty alleviation and the economic emancipation under the umbrella of ICT for development (ICT4D) domain, it failed to yield the expected results in spite of all the hype and popular rhetoric. After a decade of ICT4D initiatives, we are now at a crossroad to reflect back on ICT and smart systems' roles and contributions to the societies in developing countries from a holistic point of view. The talk will generate critical questions and comparisons with some examples from ongoing research to explore some real prospects while identifying some fads and traps.

Dr. Wei-Chung Teng

National Taiwan University of Science and Technology

Hack into The Speeds of Remote Clocks - A Promising Physical Unclonable Function for Device Identification

A smart city in the era of IoT, or Internet of things, should be full of embedded systems supporting wireless communication. These small devices are equipped with multiple sensors but limited computation power. To make sure that the traffic comes from the original neighbor or remote devices, but not from faked ones by malicious attackers, a direct answer is to adopt technologies from cryptography. However, the computation resource required for encryption/decryption and the accompanying key management protocol might become a burden to system operation. On the other hand,

IOP Conf. Series: Materials Science and Engineering **325** (2018) 011001 doi:10.1088/1757-899X/325/1/011001

the puf (physical unclonable function) is a attractive alternative to ease the needs of device identification. For every digital clock, there exists tiny error of ticking rate comparing to that of the standard time. The error in each clock, or the clock skew, is considered stable in normal temperature, and distinguishable from other clocks. Therefore, a fast measurement to clock skews of remote devices over networks helps not only to time synchronization but also device identification. In this talk I will introduce several approaches to measure the clock skews of linking devices by only few minutes, yet the precision reaches at least 0.1 ppm (parts per million). The challenges of fast measurement on multi-route network communication will also be addressed.

IOP Conf. Series: Materials Science and Engineering **325** (2018) 011002 doi:10.1088/1757-899X/325/1/011002

Peer review statement

All papers published in this volume of *IOP Conference Series: Materials Science and Engineering* have been peer reviewed through processes administered by the proceedings Editors. Reviews were conducted by expert referees to the professional and scientific standards expected of a proceedings journal published by IOP Publishing.

IOP Conf. Series: Materials Science and Engineering **325** (2018) 012001 doi:10.1088/1757-899X/325/1/012001

Mining the National Career Assessment Examination Result Using Clustering Algorithm

M V Pagudpud[1], T T Palaoag[2] and L M Padirayon[3]

1 Quirino State University, Quirino, Philippines

2 University of the Cordilleras, Baguio City, Philippines

3 Cagayan State University, Cagayan, Philippines

mvpagudpud@gmail.com

Abstract. Education is an essential process today which elicits authorities to discover and establish innovative strategies for educational improvement. This study applied data mining using clustering technique for knowledge extraction from the National Career Assessment Examination (NCAE) result in the Division of Quirino. The NCAE is an examination given to all grade 9 students in the Philippines to assess their aptitudes in the different domains. Clustering the students is helpful in identifying students' learning considerations. With the use of the RapidMiner tool, clustering algorithms such as Density-Based Spatial Clustering of Applications with Noise (DBSCAN), k-means, k-medoid, expectation maximization clustering, and support vector clustering algorithms were analyzed. The silhouette indexes of the said clustering algorithms were compared, and the result showed that the k-means algorithm with k = 3 and silhouette index equal to 0.196 is the most appropriate clustering algorithm to group the students. Three groups were formed having 477 students in the determined group (cluster 0), 310 proficient students (cluster 1) and 396 developing students (cluster 2). The data mining technique used in this study is essential in extracting useful information from the NCAE result to better understand the abilities of students which in turn is a good basis for adopting teaching strategies.

1. Introduction

Education is a fundamental sector in any nation [1] for it improves the productivity and quality of peoples' lives [2]. Furthermore, education plays a crucial role in securing national economic and social growth by equipping people with the relevant skills to fortify competitiveness and to prevent unemployment [3]. For these reasons, authorities continue to discover and proposed possibilities to strengthen education by exploring and understanding the information gathered from the educational environment.

Educational data mining (EDM) is concerned with developing methods for exploring the unique types of data that come from the educational setting and using those methods to better understand students and the settings which they learn in [4]. Within data mining, clustering is the most popular technique, and in EDM, clustering has been used in various contexts [5].

Clustering is an unsupervised learning technique which the main goal is to determine the intrinsic grouping in a set of unlabeled data using an algorithm [6][7]. It is the task of assigning a set of objects into clusters so that the objects in the same cluster are more similar to each other than to those in other clusters [8]. The clustering problem is defined as a problem of classifying a group of data points into some clusters without any prior knowledge about data structure, to produce a concise representation of the data [9]. The method is vital in data mining, statistical data analysis machine learning, pattern recognition, image analysis, information retrieval, and bioinformatics [10]. Moreover, the application of various clustering algorithm has been applied in many cases to the educational dataset in diverse studies [11].

Furthermore, clustering students is one of the frequent tasks in educational data mining. The objective is to create groups of students according to their customized features, and personal characteristics [12] and the clusters of students obtained can be used by the education administration to build a personalized learning system [13].

In the Philippines, the Department of Education conducts an annual National Career Assessment Examination (NCAE) nationwide that produces ample data set. To this day, there are hardly any researchers conducted about the result of NCAE and since the Philippine educational system tries to apply educational innovation to promote national development and global competitiveness [14], this research tries to discover novel and potentially useful information from the NCAE data.

There are eleven abilities measured by the NCAE. The scientific ability, reading comprehension, verbal ability, mathematical ability, logical reasoning ability, clerical ability, non-verbal ability, visual manipulative skill, Humanities and Social Sciences Science (HUMMS) ability, Science and Technology, Engineering and Math (STEM) ability and Accountancy, Business and Management (ABM) ability. The examination is intended to assess the preferred track of the students basing from their abilities on the different domains tested which will guide the students in choosing the track they are going to enroll in their senior high school. Since NCAE is a good test for aptitude, the result of the said examination can be a good basis to help students enhance their different abilities on the various domain tested. Hence, the conduct of this undertaking is necessary.

The main objective of this study is to cluster students based on the result of their NCAE to recommend groups of students that would lead to the general improvement of their abilities in the different areas tested by the said examination. To be able to obtain the best group for the students, clustering algorithm using the RapidMiner tool was used. The dataset was tested with an internal validation measure to select which among Density-Based Spatial Clustering of Applications with Noise (DBSCAN), k-means, k-medoid, expectation maximization clustering, and support vector clustering is the most suited algorithm for the data set. The clustering model produced by the appropriate clustering algorithm can be a basic resource in identifying students' learning considerations in the division of Quirino.

2. Methodology

The methodology used in this study is described in this section.

2.1. Data Set

The data used in this study is the NCAE 2015-2016 result obtained from the division of Quirino. The data contains 1183 instances with 11 attributes. The following are the attributes included:

- Scientific Ability (SA)
- Reading Comprehension (RC)
- Verbal Ability (VA)
- Mathematical Ability (MA)
- Logical Reasoning Ability (LRA)
- Clerical Ability (CA)
- Non-verbal Ability (NVA)
- Visual Manipulative Skills (VMS)
- Humanities and Social Sciences Science (HUMMS) Aptitude
- Science and Technology, Engineering and Math (STEM) Aptitude
- Accountancy, Business, and Management (ABM) Aptitude

They are real values ranging from 1 to 100 as shown in table 1.

Table 1. Sample NCAE data.

	SA	RC	VA	MA	LRA	CA	NVA	VMS	HUMMS	STEM	ADM
1	80	66	14	36	36	82	54	93	85	36	54
2	60	18	51	43	83	38	86	88	36	48	35
3	60	79	62	51	11	87	95	83	56	80	60

2.2. *Clustering*

Clustering is one of the techniques often used in analyzing data sets where each cluster formed is a collection of data objects that are similar to another place within the same cluster but dissimilar to objects in other clusters [15]. This study makes use of clustering analysis to segment students into groups according to their aptitude scores in the different domains tested by the NCAE.

There are various existing clustering algorithms which may all produce different segments on the same set of data. In this paper, the clustering algorithms are evaluated using internal validity measures to choose which of the algorithms is most suitable for the NCAE dataset. Five clustering algorithm were subjected to the internal evaluation which includes DBSCAN, k-means, k-medoids, expectation maximization (EM) clustering and support vector clustering (SVC). The RapidMiner tool was used to carry out clustering and internal validation.

2.3. *Internal Validation Measure*

Internal validation measure was used to choose the best clustering algorithm and the optimal cluster number without any external information [16]. In this study, the average silhouette width of DBSCAN, k-means, k-medoids, expectation maximization and support vector clustering algorithms were compared. The following is the process of computing the silhouette index [17]:

For a given cluster, Xj (j = 1,..c), the silhouette technique assigns to the ith sample of Xj a quality measure, $s(i) = (i = 1, \ldots m)$, known as the silhouette width. This value is a confidence indicator on the membership of the ith sample in the cluster Xj and it is defined as:

$$s(i) = \frac{(b(i)-a(i))}{max\{a(i),b(i)\}} \tag{1}$$

where $a(i)$ is the average distance between the ith sample and all of samples included in Xj; $b(i)$ is the minimum average distance between the $it\ sh$ and all of the samples clustered in $Xk(k = 1, \ldots c;\ k \neq j)$.

The silhouette ranges from -1 to 1, where a high value indicates that the object is well matched to its cluster and poorly matched to neighboring clusters. If the silhouette has a high value, then the clustering configuration is appropriate.

3. Results and Discussion

After executing the process, the clustering algorithms produced clustering models and average silhouette values as shown in table 2. The DBSCAN clustering produced three clusters, but its silhouette index is -0.063. A negative silhouette index is undesirable. The K-means algorithm where k = 3 has 0.196 silhouette index from three clusters. K-medoid (k=3) and EM have 0.178 and 0.090 silhouette indexes respectively, and SVC produced two clusters with 0.093 silhouette index. The silhouette indices computed were near 0 which indicates that some of the samples are very close to the decision boundary between two neighboring clusters.

Nevertheless, since the larger value of the silhouette index indicates a better clustering model, we can conclude that K-means provided the best model as compared to the other clustering algorithms tested. Cluster 0, cluster 1 and cluster 2 produced by the k-means algorithm were composed of 477, 310 and 396 students respectively.

The centroid plot produced by the k-means algorithm, where k = 3 is shown in figure 1. The centroid plot is used to inspect the clustering results. It can be inferred from the figure that the students who belong to Cluster 0 (blue line) have average scores in almost all of the abilities tested as compared to the other two clusters. This group is called the 'determined.' They scored best on the clerical ability and non-verbal ability tests, but they scored low in mathematical ability and logical reasoning ability tests.

Table 2. The result of the process.

Clustering Algorithm	Clustering Model			Silhouette Index
	Cluster 0	*Cluster 1*	*Cluster 2*	
DBSCAN	1051	115	17	-0.063
K-Means (k = 3)	477	310	396	0.196
KMedoid (k = 3)	241	414	528	0.178
SVC	373	810	0	0.093
EM (k = 3)	788	151	244	0.090

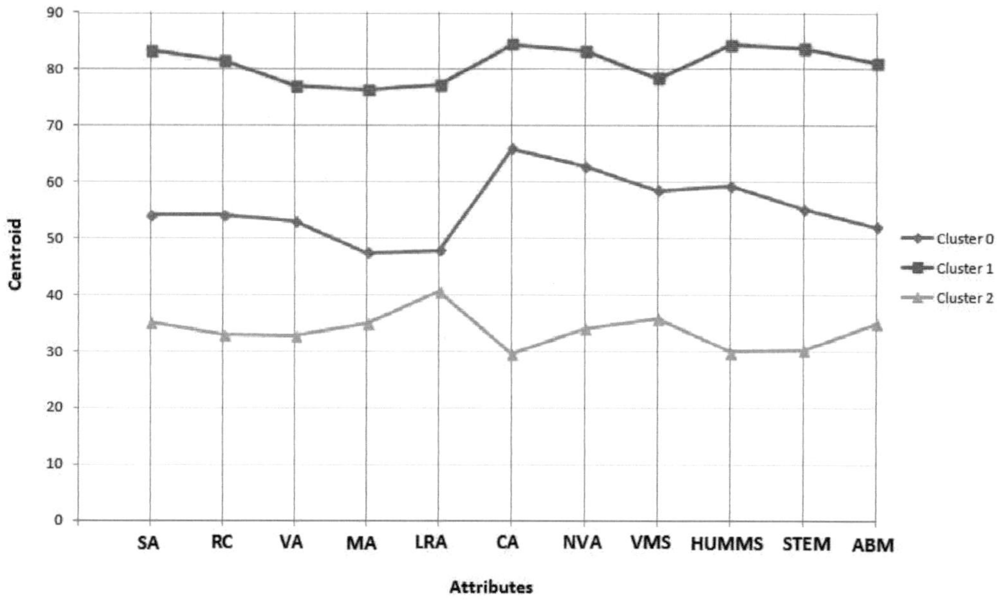

Figure 1. K-means Clustering Centroid Plot

Moreover, students who belong to cluster 1 (red line) had the best performances in all the abilities tested compared to the other clusters. Their group is called the 'proficient.' On the other hand, cluster 2 (green line) represents the students who had worst performances as compared to the other clusters. These students achieved the lowest result on clerical ability, HUMSS ability, and STEM ability tests. These students belong to the 'developing' cluster.

With this result, it would be better to apply separate learning strategies to the different clusters of students. Table 3 presents the summary of the result of the clustering technique and the students' learning considerations recommended.

4. Conclusion

This study focuses on data mining by implementing clustering algorithm on the NCAE dataset using RapidMiner tool to come out with the clusters of students. By applying an internal validation measure which is the silhouette index measure, the result showed that the K-means algorithm, using k = 3, provided the best clustering model to cluster students having a value of 0.196. Three groups were formed having 477 students in the determined group (cluster 0), 310 proficient students (cluster 1) and 396 developing students (cluster 2). The groups presented in the clustering model was produced using the k-means clustering algorithm can be used by school administration in promoting students collaborative learning, effective group training and personalized learning system in the division of Quirino. More importantly, the data mining technique used was essential in extracting valuable information from the NCAE result which is a good basis for building a learning system which can supplement student's learning. It is recommended that students' learning should consist of various learning content structure, more presentations, simulations, and exercises and should contain a reward mechanism for accomplishments to encourage student engagement.

Table 3. Summary of the clustering result and the students' learning considerations

Cluster	Clustering Technique Result	Learning Considerations
Cluster 0 *Determined Group*	- High in clerical - High in non-verbal ability - Low in mathematical ability - Low in logical reasoning - Average in other abilities	- More learning content structure in mathematics & logical reasoning - Extra learning content structure in clerical & non-verbal abilities - Supplementary learning content structure in other abilities - reward accomplishments
Cluster 1 *Proficient Group*	- Very high in Scientific Ability - Very high in Clerical Ability - Very high in HUMMS - High in others abilities	- Supplementary learning content structure in other abilities - reward accomplishments
Cluster 2 *Developing Group*	- Lowest in clerical ability - Lowest in HUMSS ability - Lowest in STEM ability - Low in other abilities	- More learning content structure in all areas - More presentations, simulations, and exercises learning contents in all areas - reward accomplishments

For future works, the researchers intend to gather additional data on recent NCAE results to be able to build a model for forecasting students' performance in the NCAE in the division of Quirino; and continue to pursue researches in data mining as applied to the educational context.

Acknowledgement
The authors extend their gratitude to the Quirino State University and the Department of Education, Division Office of Quirino for their support to the study providing all the necessary resources.

References

[1] Orji K E and M. Job 2013 The role of education in national development: Nigerian experience *European Scientific J.* **9(28)**.

[2] Taiwo R O A and Tunde A J 2016 Impact of education on national development *Int. J. of Advanced. Academic Research* **2(1)** (Social Sciences and Education).

[3] Robert A T 2015 *Relevance of Vocational and Technical Education* [Online] Available: http://www.ghanaweb.com/GhanaHomePage/NewsArchive/Relevance-of-vocational-and-technical-education-353126. [Accessed 20 April 2017].

[4] Baker R S d 2010 Mining data for student models *Advances in Intelligent Tutoring Systems* (Springer Berlin Heidelberg) pp 323-37.

[5] Mohamad S K and Tasir Z 2013 Educational data mining: a review *The 9th Int. Conf. on Cognitive Science* (Malaysia).

IOP Conf. Series: Materials Science and Engineering **325** (2018) 012001 doi:10.1088/1757-899X/325/1/012001

[6] IndiraPriya P and Ghosh D K 2013 A survey on different clustering algorithms in data *Int. J. of Modern Eng. Research* **3(1)** pp. 267-74.

[7] Feng Y 2012 Analysis on algorithm and application of cluster in data mining *J. of Theoretical and Applied Inform. Technol.* **46(1)**.

[8] Sharma N, Bajpai A and Lito R 2012 Comparison of the various clustering algorithms of weka *Int. J. of Emerging Technol. and Advanced Eng.* **2(5)**.

[9] Roseline R, Jenitha G and Amirhtaraj H 2014 Analysis and application of clustering techniques in data mining *Int. J. of Computing Algorithm* **03** pp. 910-2.

[10] Prasad J R 2014 Pattern recognition: possible research areas and issues *Int. J. of Comp. Sci.and Network* **3(5)**.

[11] Dutt A, Aghabozrgi S, Ismail M A B and Mahroeian H 2015 Clustering algorithms applied in educational data mining *Int. J. of Information and Electronics Eng.* **5(2)**.

[12] Vukićević M, Kirchner K, Delibašić B, Jovanović M, Ruhland J and Suknović M 2013 Finding best algorithmic components for clustering microarray data *Knowledge and Information Syst.* **35(1)** pp 111-30.

[13] Goyal M and Vohra R 2012 Applications of data mining in higher education *Int. J. of Computer Sci. Issues* **9(2)**.

[14] Escudero A G 2013 House Bill No.2204 [Online] Available: http://www.congress.gov.ph/legisdocs/basic_16/HB02204.pdf. [Accessed 20 April 2017].

[15] Veeramuthu P, Periyasamy R and Sugasini V 2014 Analysis of student result using clustering *Int. J. of Computer Sci. and Information Technol.* **5(4)**.

[16] Liu Y, Li Z, Xiong H, Wu J, Gao X and Wu S 2013 Understanding and enhancement of internal clustering validation measures *IEEE Transactions On Cybernetics* **43(3)**.

[17] Rendon E, Abundez I M, Gutierrez C, Diaz S Z, Arizmendi A, Quiroz E and Arzate E 2011 A comparison of internal and external cluster validation indexes *Applications of Math. and Computer Eng.*

Data Exploration and Analysis of Alternative Learning System Accreditation and Equivalency Test Result Using Data Mining

J A Talingdan[1], J T Trinidad, Jr.[2] and T D Palaoag[3]

[1] Abra State Institute of Sciences and Technology, Lagangilang, Abra

[2] Abra State Institute of Sciences and Technology, Lagangilang, Abra

[3] University of the Cordilleras, Baguio City, Philippines

janelyn.ambre@gmail.com, jun_trinidad@ymail.com, tpalaoag@gmail.com

Abstract. Alternative Learning System (ALS) is a subsystem of Depatment of Education (DepEd) that serves as an option of learners who cannot afford to go in a formal education. The research focuses on the data exploration and analysis of ALS accreditation and equivalency test result using data mining. The ALS 2014 to 2016 A & E test results in the secondary level were used as data sets in the study. The A & E test results revealed that the passing rate is doubled per year. The results were clustered using k- means clustering algorithm and they were grouped into good, medium, and low standard learners to identify students need exceptional stuff for enhancement. From the clustered data, it was found out that the strand they are weak in is strand 4 which is the Development of Self and a Sense of Community with a general average of 84.23. It also revealed that the essay type of exam got the lowest score with a general average of 2.14 compared to the multiple type of exam that covers the five learning strands. Furthermore, decision tree and naive bayes were also employed in the study to predict the performance of the learners in the A & E test and determine which is better to use for prediction. It was concluded that naive bayes performs better because the accuracy rate is higher than the decision tree algorithm.

1. Introduction

Data mining in education is called educational data mining. Educational Data Mining (EDM) is the development of methods for the exploration of data in the educational environment. It is an interesting area of research that provides better understanding on the performance of the students to improve quality of education. This can be processed through different data mining techniques such as clustering and classification [1].

Clustering is one of the data mining techniques to group the data into different clusters. It can be a data modeling technique to summarize big data. Clustering is very helpful to any disciplines and has an important role in numerous applications [2]. On the other hand, classification is a supervised learning technique to create a model to classify data according to a predefined class label. The aim of classification is for prediction to help in decision making. It can be used to predict students' performances [3].

There are studies that use educational data mining particularly the clustering algorithm to assess the performance of students and group them to determine what kind of learning strategies and techniques will be given according to the level of their understanding. This is necessary for the improvement of educational learning process. Classification is also used to predict the performance of students. This is essential to educators because it will help them in the decision making to make their students succeed. [4][5][6]

From the studies mentioned above, it is clear and understandable that educational data mining has a great impact in educational institutions. The results serve as the basis in decision making of the institution to improve the quality of education rendered to the students. Likewise in the informal basic form of education, learners' academic performance is also their main goal. It is their goal to improve the performance of learners in the competency assessment conducted every year.

In the Philippines, there is the program of the government to bring education to all Filipinos who do not have a chance to attend the formal basic education due to many reasons such as drop outs from school, no school in the communities and poverty. The program is called Alternative Learning System (ALS). The Alternative Learning System is a parallel learning system in the Philippines that is implemented by the Department of Education. It is an option of learners who cannot afford to go in a formal education. This kind of education happens outside the classroom, in community-based organizations or any areas where learning process is possible. The learners can take the Alternative Learning System Accreditation and Equivalency Test after graduation. The test covers the five learning strands namely, Communication Skills, Problem-Solving and Critical Thinking, Sustainable Use of Resources/Productivity, Development of Self, and a Sense of Community and Expanding One's World Vision. It is a paper-and-pencil test designed to measure the competencies of the ALS learners. The test is divided into two parts, the first part is the multiple choice type of exam that covers the five learning strands and the second part is the composition or the essay. Passers are given a certificate/diploma, certifying their competencies as comparable graduates of the formal school system. Passers are qualified to enroll in secondary and post-secondary schools.

In the 2016 A & E test result in the Cordillera Administrative Region, the Department of Education – Alternative Learning System Baguio Division again registered the highest passing rate in the region with an average of 82.78 percent for the elementary and secondary levels. Ifugao was second with a passing rate of 62.44 percent. Mountain Province followed third with 58.92 percent. Abra and Benguet followed closely with 57.58 percent and 57.16 percent, respectively, then Apayao at 43.05 percent, and Kalinga at 27.25 percent. It has been observed that the percentage of passers in the province of Abra is average compared to the passers of Baguio City [7]. Therefore, it is a challenge for every ALS facilitator to improve the performance of the learners in the A & E test in the province.

The research focused on Educational Data Mining (EDM) using the A & E test results of secondary level of the previous three years from 2014 to 2016 as data sets. The A & E test results were clustered using K-means algorithm of Rapid Miner to determine how the group of learners performed in the A & E test for three years. The research also determined what strands the learners are weak in and needs improvement. Furthermore, the research predicted the learners' performance in the next A & E test using the decision tree and naïve bayes algorithm. Lastly, the performance of the two algorithms was tested to determine which is better in prediction.

The result of the study will be a great help to the ALS facilitators of Abra to come up with solutions to improve the passing rate in the province. It could be beneficial by providing additional support in the delivery of instruction and improving their teaching strategies to meet the standard of learners according to the group they belong to. Through this the ALS facilitators can enhance the quality of education and provide better educational services. This will also help the ALS administration to assess, evaluate, plan and decide on how they will level up their performance in the A & E test.

1.1. Conceptual Framework

The ALS 2014 to 2016 Accreditation and Equivalency Test results who enrolled in the secondary level were used as data sets. These data sets were clustered using k-means clustering algorithm. Afterwards, the learners were grouped into three major categories as good, medium and low standard learners [5]. From the clustered data it determined the performance of the learners in the 2014 to 2016 A & E test and determined the strand(s) the learners are weak in and need improvement. On the other hand, decision tree and naive bayes were also employed in the study to predict the performance of the learners on the A & E test. It also determined which is better to use in prediction by testing their performance rate.

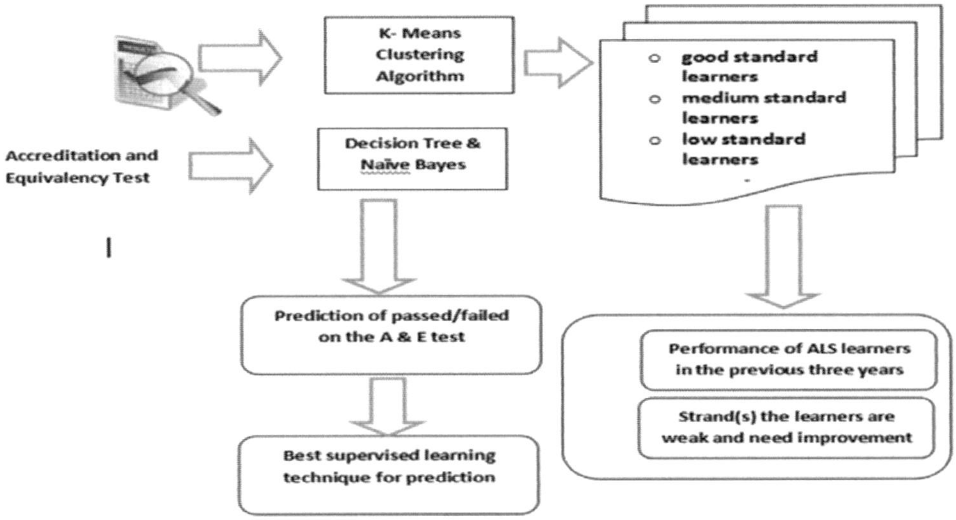

Figure 1. Conceptual Framework

2. Methodology

2.1. Data Set

The data set used in this study was the result of the A & E test of ALS learners who enrolled secondary level in the province of Abra for the past three years (2014-2016). The data is composed of six attributes namely, Communication Skills that denoted by SS1, Problem-Solving and Critical Thinking denoted by SS2, Sustainable Use of Resources/Productivity denoted by SS3, Development of Self, and a Sense of Community denoted by SS4 and Expanding One's World Vision denoted by SS5 and Essay. Table 1 presents the attributes and their values that exist in the data set as taken from the A & E test result.

Table 1. Sample A & E Test Result

	SS1	SS2	SS3	SS4	SS5	Essay	Remarks
1	103	110	114	103	120	4	Passed
2	100	110	110	102	120	3	Passed
3	93	95	105	87	105	2	Failed
4	103	110	114	99	120	4	Passed

2.2. Data Processing Models

The clustering method applied in this study was the k-means algorithm. This algorithm was used to choose the best cluster center to be the centroid. It is one of the simplest unsupervised learning algorithms used for clustering. Initially the learners are all in one group then the researchers segmented the learners into three groups based on their standard score for each strand. The clustered groups were labeled as good, medium and poor standard learner. Through this, it is easy to determine the strands that need improvement and the group of learners that need extra effort in the teaching process to improve their A & E test result.

Decision tree algorithm and naïve bayes were applied in the A & E test results to generate a model to predict learners' performance. Through these algorithms, they predicted learners who will fail and pass the A & E test. The results obtained may be helpful to the facilitators so that appropriate actions could be done to increase the passing rate. The facilitators may prepare a remedial program or more additional works for the learners predicted to fail in order for them to excel in the next A & E test.

3. Results and Discussion

From the 2014 to 2016 A & E test results of ALS learners enrolled in the secondary level that served as data sets in the study, it is noted that the passing rate is improving every year. From 11.5% of 2014, it increased to 26.5% in 2015 and jumped to 53.9% in 2016. The result signifies that there is a great improvement in the passing rate since it was doubled every year. However, the passing rate is not yet enough to make the province top the A & test compared to Baguio City which topped the 2016 A & E test with a passing rate of 82.78. Table 2 shows the percentage rate of those who failed and passed.

Table 2. Percentage rate of passed and failed learners

Remarks	2014		2015		2016	
	No of learners	**Percentage**	**No of learners**	**Percentage**	**No of learners**	**Percentage**
Passed	45	11.5	122	26.5	219	53.9
Failed	347	88.5	339	73.5	187	46.1
Total	392	100%	461	100%	406	100%

3.1. Cluster Model

From the 2014 to 2016 A & E test results, k – means clustering was applied and group the learners into three clusters. For year 2014, there were 392 learners who took the A & E test and 91 of them were grouped in cluster 0, 148 learners for cluster 1 and 153 learners to cluster 2. For year 2015, there were 221 learners grouped to cluster 0, 1 leaner to cluster 1 and 239 learners to cluster 2 to give a total of 461 learners who took the A & E test. Lastly, for year 2016 there were 139 learners grouped in cluster 0, 93 learners in cluster 1 and 174 learners in cluster 2 to make the 406 learners who took the A & E test on that year. Table 3 shows the distribution of the learners.

Table 3. Distribution of learners

Cluster	2014	2015	2016
Cluster 0	91	221	139
Cluster 1	148	1	93
Cluster 2	153	239	174
Total	392	461	406

3.2. Centroid Table

Table 4 is the centroid table and it is observed that from 2014 to 2016, the essay type of exam is the weakest area of the learners with a general average of 2.14 compared to the multiple choice type of exam that covers the five learning strands. From the five learning strands, strand 4 which is the Development of Self and a Sense of Community is the strand that the learners are weak in and needs improvement because it has the lowest score of a general average of 84.23. Moreover, learners should be given different teaching strategies and learning techniques based on the group they belong. Learners labeled as poor standard learners should be given exceptional stuff for enhancement compared to other groups as they got low performance in all strands. For the good standard learners, the strand that they got low score should be given extra emphasis to help them excel in the A & E test. Lastly, the average standard learners should be given more attention like giving extra work especially on the strands that they are weak in to improve their score in the A & E test. Along with this, more exercises should be given on essay type of exam to improve the test result.

From the k-means clustering centroid table of 2014 A & E test result, it can be perceived that in the multiple choice type of exam, learners belong to cluster 2 had mediate scores in all strands compared to the other two clusters. This group is called medium standard learners. They scored best in strand 5 which is the Expanding One's World Vision but they score low in strand 4 which is the Development of Self and a Sense of Community. Learners who belong to cluster 1 had the best performance in all aspects of exam compared to other clusters. Their group can be grouped as good standard learners. On the other hand, cluster 0 got the lowest score and can be considered as poor standard learners. It can be observed also that all clusters did not excel in the essay type of exam.

Table 4. Centroid Table

Attri-bute	2014 Clusters				2015 Clusters				2016 Clusters				Gen. AVG.
	0	1	2	AVG.	0	1	2	AVG.	0	1	2	AVG.	
SS1	78.64	93.66	87.18	86.49	93.03	98	83.85	91.63	98.89	83.88	91.23	91.33	89.82
SS2	87.37	98.76	93.65	93.26	98.98	95	91.16	95.05	105.4	90.86	96.78	97.68	95.33
SS3	80.18	96.89	87.65	88.24	97.3	95	85.09	92.46	109.1	83.93	91.53	94.86	91.85
SS4	72.07	88.85	80.85	80.59	88.22	88	78.75	84.99	97.62	77.52	86.2	87.11	84.23
SS5	90.53	115.7	100.9	102.4	115.3	14	97.65	75.65	118.8	96.47	113.4	109.5	95.86
Essay	1.275	1.905	1.641	1.607	2.715	3	2.188	2.634	1.115	2.484	2.931	2.177	2.14

In addition, It can be also inferred in the centroid table of 2015 that in the multiple choice type of exam, learners belong to Cluster 1 had mediate scores and they excel more in strand 1 which is Communication Skills compared to other clusters but they got a very low performance in strand 5 which is Expanding One's World Vision. This group is called average standard learners. Learners who belong to cluster 0 had the best performance in four strands except strand 1 which is lower compared to cluster 1. Learners who belong to this group are called good standard learners. On the other hand, cluster 2 represents learners who had worst performance compared to the two clusters and can be labelled as poor standard learners. It is also noted that all clusters did not perform well in the essay type of exam.

Furthermore, from the centroid table of A & E test result of 2016 it shows that learners who belong to cluster 0 had the highest score and can be labelled as good standard learners. Cluster 1 is the least performing group and can be labelled as poor standard learners and the learners who belong to cluster 2 can be labelled as average standard learners. It can also be observed that in the multiple choice type of exam all clusters excelled in strand 5 which is the Sense of Community and Expanding One's World Vision but they all got low score in strand 4 which is the Development of Self. Essay type of exam is still the hardest part of the exam as all clusters got low score compared to the multiple type of exam that covers the five learning strands.

3.3. Prediction

Naïve Bayes and Decision tree algorithms were used to predict the performance of learners on the 2016 A & E test. The 2014 and 2015 A & E test results were used as the training data set in the study. The performance rate of the two supervised learning techniques was tested to determine which is better to use in predicting the performance of learners. It has been noted that the Naïve Bayes is more accurate than the decision tree as the accuracy rate is 94.83% compared to 87.44% accuracy rate of decision tree. When it comes to precision, decision tree is more precise with 97.73% compared to 91.25% precision rate of Naïve Bayes but in the recall it can be observed that Naïve Bayes got a perfect rate of 100% compared to the 78.54% of decision tree. From this performance test, it shows that naïve bayes is better to use when it comes to the prediction of performance of the ALS learners in the A & E test compared to the decision tree. Prediction of performance of learners is necessary in order for the facilitators to identify the learning techniques and teaching strategies they will use to improve the learners' performance. This is also to help the facilitators to determine learners that need more attention for them to improve.

The result of the study supported the research entitled Educational Data Mining to Reduce Student Droput Rate by Using Classification wherein the aimed was to evaluate student performance by using different decision tree and bayes algorithms. This help them to identify the weak students having enrollment status at risk and students needing further help. It was found out that naive bayes is the best algorithm as it gets the highest accuracy rate which is above ninety percent compared to other algorithms used in the study. Table 5 below is the performance rate of the two algorithms [8].

Table 5. Performance Rate

	Naïve Bayes	Decision Tree
Accuracy	94.83%	87.44%
Precision	91.25%	97.73 %
Recall	100%	78.54%

IOP Conf. Series: Materials Science and Engineering **325** (2018) 012002 doi:10.1088/1757-899X/325/1/012002

4. Conclusion

The percentage of passers significantly increased every year as a new set of A & E takers (ALS learners) takes the exam from 2014 to 2016. This is a reflection of some form of intervention by the ALS facilitators because teacher factor cannot be undervalued. Considering that the results on essay being consistently the attribute in which most learners fail every year, this could be associated with their weak foundation in reading and writing. In the same manner that strand 4 or Development of Self and a Sense of Community is also consistent in being the strand in which the learners performed the least among the strands alone excluding essay. This may be associated to their lack of self and social awareness and participation. Naïve Bayes being the better supervised learning technique for prediction as tested in this study may be linked with the fact that it is based on the theory of probability and not on the divide and conquer technique as being used in the decision tree. Thus, the exploration of ALS A & E test results is very crucial in determining the performance of ALS learner on the exam so proper application of data mining techniques shall be used to achieve enhancement in the exam.

References

[1] Romero C and Ventura S 2013 Data mining in education *Wiley Interdisciplinary Reviews: Data Mining and Knowledge Discovery* 3(1) pp 12-27
[2] Berkhin P 2006 A Survey of clustering data mining techniques *In Grouping Multidimensional Data* (Heidelberg: Springer Berlin) pp 25-71.
[3] Al-Barrak M A and Al-Razgan M S 2015 Predicting students' performance through classification: a case study *J. of Theoretical & Applied Information Technol.* **75(2)**.
[4] Oyelade O J, Oladipupo O O and Obagbuwa I C 2010 Application of k means clustering algorithm for prediction of students academic performance *arXiv* (*preprint* arXiv:1002.2425).
[5] Shovon M, Islam H and Haque M 2012 An approach of improving students academic performance by using k means clustering algorithm and decision tree *arXiv* (*preprint* arXiv:1211.6340).
[6] Baradwaj B K and Pal S 2012 Mining educational data to analyze students' performance *arXiv* (*preprint* arXiv:1201.3417).
[7] Dar C 2016 Baguio City posts 82% passing rate in 2016 A & E exam.
[8] Shakeel K and Butt N A *Educational data mining to reduce student droput rate by using classification.*

IOP Conf. Series: Materials Science and Engineering **325** (2018) 012003 doi:10.1088/1757-899X/325/1/012003

Validating YouTube Factors Affecting Learning Performance

Yoga Pratama[*)], Rudy Hartanto[**)], Sri Suning Kusumawardani [***)]

Department of Electrical Engineering and Information Technology, Universitas Gadjah Mada

[*]yoga.cio15@mail.ugm.ac.id, [**] rudy@ugm.ac.id, [***] suning@mti.ugm.ac.id

Abstract. YouTube is often used as a companion medium or a learning supplement. One of the educational places that often uses is Jogja Audio School (JAS) which focuses on music production education. Music production is a difficult material to learn, especially at the audio mastering. With tutorial contents from YouTube, students find it easier to learn and understand audio mastering and improved their learning performance. This study aims to validate the role of YouTube as a medium of learning in improving student's learning performance by looking at the factors that affect student learning performance. The sample involves 100 respondents from JAS at audio mastering level. The results showed that student learning performance increases seen from factors that have a significant influence of motivation, instructional content, and YouTube usefulness. Overall findings suggest that YouTube has a important role to student learning performance in music production education and as an innovative and efficient learning medium.

1. Introduction

Online material can be easily obtained from the internet, one of the providers of such material is YouTube. It allows teachers to include online material into traditional classes [1]. The lecturers provide the supreme material and the main material in the teaching process [2], [3]. The purpose of using technology in the learning process is to improve the quality of learning [4]. Jogja Audio School (JAS) is a music education institution focused on music production. Studying the music production process is not easy, especially at the audio mastering level, the material that is studied quite a lot and complicated, so that students often find it difficult to learn and understand the audio mastering process. In the process of learning JAS has developed a learning method by utilizing YouTube as a learning medium or as a student learning supplement. After using YouTube, students' learning performance is increasing. This shows that after the use of YouTube a positive change in student learning performance. In previous studies, YouTube has shown a positive impact on students' learning outcomes [5], [6], [7]. However, until now there is still no research on the factors that impact on improving student learning performance by using YouTube video tutorials to validate that YouTube has a major role in the improvement. This is the focus of this research, by analyzing the various factors that influence the improvement of student learning performance by utilizing YouTube as a medium and learning supplement. The final result will be obtained by several factors that can impact on the improvement of learning performance, and can be concluded from the conclusion is expected to give a positive impact that is the potential development of YouTube as a medium of learning in order to achieve increased learning outcomes.

2. Literatur review and hyphoteses

2.1. Motivation and Self-efficacy

In an educational environment or ordinary motivational learning is defined as "the special desire of learners to learn the contents of the learning program" [8]. In Mathieu, Tannenbaum, and Salas research, it shows that learners provide a good emotional response when they have a higher motivation. This is found proportionally with regard to improved work performance after training [9]. In Chou & Wang's research, computer self-efficacy is a very important characteristic in electronic training or e-training situations [10]. Self-efficacy also has an impact on the effectiveness of learning in such areas as in computer software learning [11]. Learning effectiveness of learners with computer self-efficacy is higher better than learners with low self-efficacy computer. Application of technology in learning will be more effective if the system has high functionality and learners have high self-efficacy computer [12]. This study intends to verify the link between student learning, motivation and YouTube self-efficacy and learning performance.

H 1-1: Learning motivation has a positive effect on learning performance
H 1-2: Learning motivation has a positive effect on the transfer of performance
H 2-1: YouTube Self-efficacy has a positive effect on learning performance
H 2-2: YouTube Self-efficacy positively affects the transfer of performance

2.2. Instructional Contents

The material of the content is certainly very important in the use or distribution of learning outcomes [13]. The material or content presented should be harmonized with actual practice, in order for an organization to increase the motivation of the trainee to improve the learning performance [70]. When learners feel the content is practicable, then they apply the ability of what they get from training in their real work [14]. When applied to the domain of this research, YouTube may be a rich and fresh source of procedural content. Eiriksdottir and Catrambone [15] stated that individuals do not see procedural or instructional content that is inaccurate and missed out as something useful. Therefore, content, or information according to Tung [16] involves three dimensions: relevance, timeliness, and sufficiency. Thus, one may demonstrate a greater perception of the benefits of a system designed to be accessible for every situational task objective. Collectively, it is expected the extent to which users feel that YouTube provides procedural learning content when requested and is perceived to affect its usefulness [17].

H 3-1: Instructional contents have a positive effect on learning performance
H 3-2: Instructional contents positively affect the transfer of performance

2.3. Usefulness and Ease of use

Perceived usefulness is defined as "the extent to which a person believes that using a particular system will improve his performance" [18]. Chiu, et al. [19] suggest that perceived usefulness positively affects the effectiveness of e-learning. Johnson et al positively demonstrate technology usefulness affecting e-learning success [20]. Based on the theory of Davis (1989) [18] shows that ease of use positively impacts perceived usefulness. Venkatesh & Davis, 1996, 2000. TAM theorizes that perceived ease of use determines the actual intent and usage behavior [21].

H 4: Ease of use has a positive effect on learning performance
H 5: YouTube usefulness has a positive effect on learning performance

2.4. Transfer Performance

Baldwin & Ford's research [22] demonstrates the integration between learning and performance transfer, learning performance influencing the transfer of performance or the use of learning outcomes in the real world. However, design, methods, materials, and theories have only an impact on learning performance. Learning performance affects the transfer of performance or application of acquired knowledge.

H 6: learning performance has a positive effect on the transfer performance

This research applies the research model from Hyochang Lim et al. [13] and there is an addition of YouTube Usefulness variable in this research to see the impact of technology in terms of expediency in the process of improving learning performance. In addition, there are some variables that are omitted because they do not fit the research needs or not in accordance with the learning culture of the Jogja Audio School, such as e-mail exchange, face-to-face, continuous learning culture and seniors' support. The following variables are used in this study:

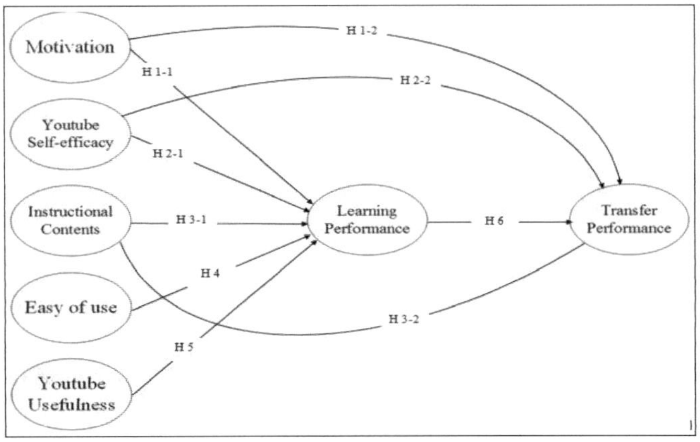

Figure 1. Proposed research model.

3. Research method

The population of this research is the students and teachers from the Jogja Audio School who have been through the learning system using video tutorial YouTube. Samples to be taken amounted to 100 respondents. Determination of the sample based on the purposive sampling method, i.e. the determination of the sample by the researcher by considering that the information needs can be obtained from the target that has been determined [23].

- The proportion of males to females is 83 percent to 27 percent.
- The age is divided into 4 groups of age ranges, age group 15-20 years as many as 10 percent, 21-25 years as many as 44 percent, 26-30 years some 39 percent and 31-35 years of 7 percent.

Primary data collected from the respondents by division of questionnaire will then be analysed. Prior to the division, according to the provisions shall be tested the validity and reliability of the instrument. The validity test is intended to assess the feasibility of the indicator or team in defining the construct variable. In the PLS, convergent validity can be seen through the loading factor of the indicators that measure the construct. The indicator can be declared valid if it has a loading factor> 0.70 [24]. While the reliability test the consistency of measuring instruments in measuring a concept correctly and error free. In PLS, reliability is tested using two methods of cronbach's alpha and composite reliability. The data is said to be good if cronbach's alpha and composite reliability have values greater than 0.7 for confirmatory research or more than 0.6 for explanatory research [24].

4. Results

The objective of this study is to identify and validate the factors that affect learning performance by using YouTube tutorial video for music production education. With the obtained results, descriptive analysis and structural equation modeling analysis were conducted.

4.1. Validity Analysis

Validity testing is done to see the suitability of each indicator in the size of a variable according to the theory used as a reference. For model models or outward models, it must look at the values of convergent validity (convergent validity) and discriminant validity (discriminant validity) of its indicators [24]. The low loading factor value is in the instructional content indicator IC6 that is equal to 0.814. Based on these results, all indicators of this research model are valid. Discriminant validity is indicated by the correlation between measuring devices in measuring different attributes, the expected results are not extreme and smaller than the measuring instruments that measure the same attributes. The discriminant validity table shows a good result based on the value of the Fornel - Larcker criterion which shows the square root of the AVE value of each latent variable is greater than the value of latent variable correlations as can be seen in Table 1.

Table 1. Discriminant validity based on Fornel-Larcker criterion.

	SE	IC	EOU	MV	LP	TP	YU
SE	**0.860**						
IC	0.863	**0.904**					
EOU	0.783	0.771	**0.909**				
MV	0.752	0.801	0.731	**0.859**			
LP	0.738	0.844	0.668	0.814	**0.920**		
TP	0.752	0.811	0.714	0.791	0.852	**0.948**	
YU	0.765	0.850	0.735	0.854	0.886	0.865	**0.904**

Note. SE = Self-Efficacy; IC = Instructional Contents; EOU = Easy of use; MV = Motivation; LP = Learning Performance; TP = Transfer Performance; YU = Youtube Usefulness.

4.2. Reliability Analysis

Reliability testing is intended to see the consistency of the measurement gauge of an instrument if the instrument is reused as a measuring instrument. In reliability, it can use Cronbach's Alpha (CA). This value reflects the reliability of each indicator in the model. The minimum value of 0.7 is ideally 0.8 or 0.9. In addition to Cronbach's Alpha also used the value of composite reliability (CR) which is interpreted the same as Cronbach's Alpha [24]. The value of CA & CR in this research model can be seen in Table 2.

Table 2. Composite Reliability (CR) and nilai Cronbach Alpha (CA).

Constructs	*Cronbach's Alpha*	*Composite Reliability*
Self-Efficacy	0.912	0.934
Intructional Contents	0.968	0.973
Easy of use	0.930	0.950
Motivation	0.881	0.919
Learning performance	0.940	0.957
Transfer performance	0.888	0.947
YouTube usefulness	0.944	0.957

The value of composite reliability (CR) and Cronbach's Alpha (CA) value for all constructs worth more than 0.6 is above the minimum value, thus all constructs are considered reliable.

4.3. Results of hypothesis verification

In this research hypothesis testing by using t-value that compared with t-table. If the value of t-value is greater than the t-table value, then the hypothesis is accepted, and rejected if the t-value is less than the value of t-table [24]. One-tailed test with a significance level of 5% or 0.05 which means that research results can be accounted for in case of error in the research process of a magnitude not more than 5%. Samples of 100 respondents with 7 research variables in the model considering degrees of freedom or degree of freedom (df) obtained t-table of 1.66. Based on the results of the significance of path coefficient testing, it can be seen testing the rejected and accepted hypothesis in this study in Table 3.

Table 3. Hypotheses testing results.

Hypotheses	Path	t-value	t-table	Adopted or denied
Hypothesis 1-1	MV →LP	1.746	1.66	Adopted
Hypothesis 1-2	MV→TP	2.222	1.66	Adopted
Hypothesis 2-1	SE→LP	0.236	1.66	Denied
Hypothesis 2-2	SE→TP	1.222	1.66	Denied
Hypothesis 3-1	IC→LP	3.362	1.66	Adopted
Hypothesis 3-2	IC→TP	0.826	1.66	Denied
Hypothesis 4	EOU→LP	1.364	1.66	Denied
Hypothesis 5	YU→LP	5.292	1.66	Adopted
Hypothesis 6	LP→TP	5.012	1.66	Adopted

First, the greater motivation to learn with YouTube better learning performance and transfer performance, this means that the willingness to learn is an important factor in improving learning performance and its use in real life or work. Second, self efficacy has no effect on learning performance and performance transfer. From the test results obtained t-statistical value of SE→LP 0.236 and SE→TP 1.222 . This is unlike the study [10] [11]. Based on interview with lecturer and respondents, the material content chosen by the teacher leaves students free and has confidence in using YouTube, even though students are given the opportunity to search for personal content, but the content is subject to certain restrictions by the teacher. Third, Instructional contents improve learning performance, based on t-statistics IC→LP 3.362 and also in accordance with previous studies [25]. The ultimate goal of introducing YouTube as learning tool is to increase the learning performance with advanced student's knowlege and tecnology. However, instructional contents do not play a significant role in transfer performance with t-statistics IC→TP 0.826. Based on the research, the relationship between the two constructs requires a mediating factor that is learning performance. Fourth, ease of use has no significant effect on learning performance. With the calculation of t-statistics EOU→PB 1,364. Based on interviews with lecturer and respondents, the increased learning performance is not influenced by the ease of using YouTube, but rather the content of learning from YouTube. Fifth, learning performance is enhanced in terms of YouTube usefulness. Students feel YouTube is very useful in supporting the learning process. This is evidenced by the acquisition of t-statistics YU→LP 5.292 and support the prior study [18]. Sixth, higher learning performance better transfer performance as shown the t-statistics of LP→TP 5.012. Improved learning performance provides students with more ability to apply learning outcomes to the real work.

5. Conclusion

Thus, judging from the viewpoint of technology or learning with the YouTube media has a significant impact on improving student learning performance in terms of tutorial content and usefulness and by providing motivation in learning more. In this study most of the hypothesis stated acceptable and show that YouTube plays an important role in improving student learning performance in music production, especially audio mastering.

References

[1] National B A K 2016 *Teaching EFL to the iGeneration : A Survey of Using YouTube as Supplementary Material with College EFL Students in Taiwan.*

[2] Liu M 2016 Internet and Higher Education Blending a class video blog to optimize student learning outcomes in higher education *Internet High. Educ.* **30** pp 44–53.

[3] Uren M, Uren J, Uren M, and Uren J 2016 *eTeaching and eLearning to enhance learning for a diverse cohort in engineering education* **52**.

[4] Fullan M 2002 El significado del cambio educativo: un cuarto de siglo de aprendizaje *Rev. currículum y Form. del Profr.* **6(1998)** pp 1–14.

[5] Burke S C, Snyder S, and Rager R C 2009 *An Assessment of Faculty Usage of YouTube as a Teaching Resource* **7(1)** pp 1–8.

[6] Oddone C 2011 *Using Videos from YouTube and Websites in the CLIL Classroom* no 18 pp 105–10.

[7] Chtouki Y, Harroud H, and Ii A H 2012 *The Impact of YouTube Videos on the Student ' s Learning* pp 1–4.

[8] Noe R A and Schmitt N 1986 The influence of trainee attitudes on training effectiveness: test of a model *Pers. Psychol.* **39(3)** pp 497–523.

[9] Mathieu J E, Tannenbaum S I, Salas E, and Mathieu J E 2014 *Influences of Individual and Situational Characteristics on Measures of Training Effectiveness State University Of New York at Albany* **35(4)** pp 828–47.

[10] Chou H W and Wang T B 2000 Influence of learning style and training method on self-efficacy and learning performance in WWW homepage design training *Int. J. Inf. Manage.* **20(6)** pp 455–72.

[11] Gist M E, Schwoerer C, and Rosen B 1989 Effects of alternative training methods on self-efficacy and performance in computer software training *J. Appl. Psychol.* **74(6)** pp 884–91.

[12] Chien T C 2012 *Computer Self-Efficacy and Factors Influencing E-Learning Effectiveness.*

[13] Lim H, Lee S, and Nam K 2007 *Validating E-Learning Factors Affecting Training Effectiveness* **27** pp 22–35.

[14] Alliger G M, Tannenbaum S I, Bennett Jr W, Traver H, and Shotland A 1997 A meta-analysis of the relations among training criteria *Pers. Psychol.* **50(2)** pp 341–58.

[15] Eiriksdottir E and Catrambone R 2011 Procedural Instructions, principles, and examples: how to structure instructions for procedural tasks to enhance performance, learning, and transfer *Hum. Factors J. Hum. Factors Ergon. Soc.* **53** pp 749–70.

[16] Yang T and Maxwell T A 2011 Information-sharing in public organizations : A literature review of interpersonal , intra-organizational and inter-organizational success factors *Gov. Inf. Q.* **28(2)** pp 164–75.

[17] Lee D Y and Lehto M R 2013 User acceptance of YouTube for procedural learning: An extension of the Technology Acceptance Model *Comput. Educ.* **61(1)** pp 193–208.

[18] Davis F 1989 Perceived usefulness, perceived east of use, and user acceptance of information technology *MIS Q.* **13(3)** pp 319–40.

[19] Chiu C M, Hsu M H, Sun S Y, Lin T C, and Sun P C 2005 Usability, quality, value and e-learning continuance decisions *Comput. Educ.* **45(4)** pp 399–416.

[20] Johnson R D, Hornik S, and Salas E 2008 An empirical examination of factors contributing to the creation of successful e-learning environments *Int. J. Hum. Comput. Stud.* **66(5)** pp 356–69.

[21] Venkatesh V and Davis F D 2000 A theoretical extension of the technology acceptance model: four longitudinal studies *Manage. Sci.* **46(2)** pp 186–205.

[22] Baldwin T T and Ford J K 1988 *Transfer of Training : A Review and Directions for Future Research.*

[23] Tongco M D C 2007 Purposive sampling as a tool for informant selection *Ethnobot. Res. Appl.* **5** pp 147–58.

[24] Hair J F, Ringle C M, and Sarstedt M 2011 PLS-SEM: Indeed a silver bullet *J. Mark. Theory Pract.* **19(2)** pp 139–52.

[25] Michigan J K, Ford S U, Ethyl S P, and Wroten C 1984 Introducing new methods for conducting training evaluation and for liking training evaluation to program redesign *Pers. aPsychol.* **37** pp 651–65.

IOP Conf. Series: Materials Science and Engineering **325** (2018) 012004 doi:10.1088/1757-899X/325/1/012004

Investigating Daily Mobile Device Use Among University Students in Indonesia

Ahmad R Pratama[1,2]

[1] Department of Technology and Society, Stony Brook University, NY 11794 USA
[2] Department of Informatics, Universitas Islam Indonesia, Sleman, DIY 55584 Indonesia

E-mail: ahmad.pratama@stonybrook.edu; ahmad.rafie@uii.ac.id

Abstract. This study investigates daily mobile device use among Indonesian university students by using survey data collected in the first quarter of 2016. The data show that 55.49% of the students spend more than five hours a day on their mobile devices. OLS regression analysis shows that female students, owners of the most expensive smartphone or tablet, and those who access social media apps and play mobile games frequently are more likely to spend more time on their mobile devices each day. Contrary to that, those who consider themselves as laggards are less likely to do so. The findings in this study help better understand how university students in Indonesia use their mobile device and lay a foundation as part of a larger research agenda about the relationship between mobile technology and education in Indonesia.

1. Background

Mobile devices have been around since the 1970s, marked by the first handheld mobile phone produced by Motorola in 1973. After few decades used solely as a communication device, for both voice-based and text-based communication, mobile phones received a major upgrade in its ability to resemble a portable personal computer with a mobile operating system, also known as smartphones. While Japanese NTT DoCoMo was among the first to release smartphones and relatively successful in achieving mass adoption within a country, it was not until the next generation of smartphone, marked by the release of Apple iPhone and Google Android OS in 2007 and 2008 respectively that smartphone became a worldwide phenomenon. With the release of iPad and multiple variants of Android counterpart in the 2010s, tablets were the latest addition to the lineup of mobile devices. The trend shows that personal computing devices are becoming more and more mobile oriented.

Meanwhile, in Indonesia, the statistics of information and communication technology (ICT) adoption are more than promising. Within the past three years, Internet penetration went up significantly from 15% in 2014 to 51% in 2017, placing Indonesia as the country with the highest growth in Internet user, more than five times of the global average between 2016 and 2017 [1-4]. The number of mobile connections in Indonesia already outnumbered the total population by 2014 (112%). In this regard, Indonesia is ahead of many developed countries like the US, France, Australia, Japan, or South Korea and this is interesting, given that digital divide has been widely accepted as a worldwide problem between developing and developed countries for the past decade [5]. Also, considering that Indonesia is the fourth most populous country in the world with roughly 260 million population, this number in growth and rate means a lot as it translates to a huge number of people. These statistics do not necessarily represent the quality of technology use in any way, but they show that Indonesian people are thriving regarding technology adoption, especially in the mobile frontline.

The relationship between mobile devices and educational attainment has been investigated by researchers for a while. Some studies indicate mobile device use can help improve education [6-9] and emphasize the opportunity of transforming education with the use of mobile learning [10-11]. Some other, however, suggest a problematic relationship between them. In many cases, it is mostly due to addiction [12-16]. The same addiction problem has also been pointed out by researchers in terms of Internet access, also known as pathological Internet use [17-19]. Nonetheless, it is also important to note that the relationship between ICT, in which mobile device plays an important role nowadays, and education is a two-way effect. A higher participation rate in post-elementary education increases ICT utilization, which in turn helps promote better education outcomes [20].

In terms of Indonesia, particularly university student population, previous research in an Indonesian university shows that more than 95% of the students own smartphones by 2016 [21]. This number is significantly higher than the ownership rate of overall population in Indonesia (21%) or even in the United States (72%), which is one of the highest in the world in the same year [22]. It is also worth to note that mobile device ownership is growing rapidly among university students all over the world, as suggested by many studies [23-25]. In terms of the time-spent accessing internet from mobile device, a global survey of internet users aged 16-64 in the second half of 2016 shows that Indonesian users spend 235 minutes each day in average. This number is significantly higher than U.S. users who spend 122 minutes each day in average. Only users in Thailand (254 minutes) and Brazil (236 minutes) spend more time using mobile internet than Indonesian users [4].

While accessing the Internet might be one of the main activities people do on their mobile device, it is not the only activity. In fact, most people spend more time on the apps instead of mobile browser on their smartphones. A market research shows that U.S. consumers time-spent on mobile devices have crossed five hours (300 minutes) a day by the end of 2016 [26]. Unfortunately, there is no comparable data for Indonesian consumers yet. To fill that gap and to lay a foundation for a larger research agenda about the relationship between mobile technology and education in Indonesia, this study aims to investigate daily mobile device use among Indonesian university students in terms of time spent on mobile device and factors influencing it.

2. Data and Methodology

An online survey was conducted in a private university in Indonesia within the first quarter of 2016. A total of 182 undergraduate students (78 females, 42.86% and 104 males, 57.14%) participated in the study. They came from 7 different majors, categorized further into science, technology, engineering, math, and medicine (STEMM) and social sciences. They ranged from 19 to 25 years of age (Mean = 21.42, SD = 1.53). Participants were also categorized based on year in college (26 1st year, 14.29%, 30 2nd year, 16.48%, 58 3rd year, 31.87%, 39 4th year, 21.43%, and 29 other in their 5th year or above, 15.93%), place of origin (116 Java, 63.74% and 66 outside Java, 36.26%), self-perceived ICT adoption level representing their attitude toward ICT in three simplified scale (38 early adopters, 20.88%, 123 majorities, 67.58%, and 21 laggards, 11.54%), and socioeconomic status (SES) measured by the purchase price of their devices. Those who own the most expensive smartphones or tablets (i.e., priced at IDR 5 million or above) are considered as high SES (35 high SES, 19.23% and 147 middle-to-low SES, 80.77%). Students were also categorized based on their use of educational apps, social media apps, and mobile gaming apps on their mobile device. For these last three categories, regular users are defined as those who use at least one app in each category at least once a week. Data analysis includes both descriptive and inferential statistics in the form of OLS regression.

For the dependent variable in this study, students were asked to self-report their daily mobile device use in hours. Table 1 provides information on students with daily mobile device use of five hours or more, categorized by gender, age, major, year in college, place of origin, socioeconomic status, attitude toward ICT, educational apps use, social media apps use, and mobile gaming apps use. All ten independent variables are included as predictors in the initial OLS regression model. Stepwise deletion method is used to get the final model with only significant predictors included.

Table 1. Descriptive statistics of students with daily mobile device use of 5 hours or more per day

	≥ 5 hours		**Base**	
	Freq	**% from Base**	**Freq**	**% from Total**
Gender: Male	49	*47.12*	104	*57.14*
Female	52	*66.67*	78	*42.86*
Age: ≤ 20	28	*57.14*	49	*26.92*
21	26	*52.00*	50	*27.47*
22	26	*60.47*	43	*23.63*
≥ 23	21	*52.50*	40	*21.98*
Major: STEMM	57	*50.89*	112	*61.54*
Social Sciences	44	*62.86*	70	*38.46*
Year in College: 1st year	13	*50.00*	26	*14.29*
2nd year	18	*60.00*	30	*16.48*
3rd year	33	*56.90*	58	*31.87*
4th year	21	*53.85*	39	*21.43*
≥ 5th year	16	*55.17*	29	*15.93*
Place of Origin: Java	65	*56.03*	116	*63.74*
Non-Java	36	*54.55*	66	*36.26*
SES: High	29	*82.86*	35	*19.23*
Middle to Low	72	*48.98*	147	*80.77*
Attitude toward ICT: Early Adopter	24	*63.16*	38	*20.88*
Majority	70	*56.91*	123	*67.58*
Laggard	7	*33.33*	21	*11.54*
Educational Apps: Regular User	87	*56.13*	155	*85.16*
Non-Regular User	14	*51.85*	27	*14.84*
Social Media Apps: Regular User	101	*57.06*	177	*97.25*
Non-Regular User	0	*0.00*	5	*2.75*
Mobile Gaming: Regular User	40	*65.57*	61	*33.52*
Non-Regular User	61	*50.41*	121	*66.48*
All Samples	101	*55.49*	182	*100.00*

3. Results and Discussions

Based on the descriptive statistics in Table 1, the percentage of students who spend 5 hours or more a day on their mobile device is 55.49% for all samples, and they range from 50% to 60% in all groups when categorized based on age, major, year in college, place of origin, and educational apps use. However, it is not the case when these students are categorized based on gender, SES, attitude toward ICT, social media use, and mobile gaming use where the percentage is as low as 0% in one group (i.e., non-regular user of social media apps) and as high as 82.86% in another (i.e. students with high SES). Significant differences in the percentage of students who spend 5 hours or more a day in these categories are confirmed by applying independent group t-test and chi-square test accordingly.

Regardless of the difference, all ten independent variables are included in the original model upon running the multivariate analysis with OLS regression. After stepwise deletion method, only five predictors are found to be significant as shown in Table 2. These five significant predictors are gender ($Pr < .05$), negative attitude toward ICT ($Pr < .05$), high SES ($Pr < .001$), regular use of social media apps ($Pr < .001$), and regular use of mobile gaming apps ($Pr < .05$). Unsurprisingly, these five are also the ones with a significant difference in the percentage of students who spend 5 hours or more on their mobile device per day. The final model uses robust standard errors due to an indication of heteroscedasticity problem in the original model based on Breusch–Pagan test while Ramsey RESET test is used to confirm no omitted-variable bias. VIF value indicates no multicollinearity issue.

Of those five significant predictors, regular use of social media has the strongest effect of all. Holding everything else constant, those who use the apps at least once a week are likely to spend 3.01 hours (about 181 minutes) more on their mobile device per day than those who don't. The second strongest predictor is high SES. Holding everything else constant, those who own the most expensive smartphone or tablet are likely to spend .663 hours (about 40 minutes) more on their mobile device per day than those who don't. The next strongest predictor is negative attitude toward ICT. Holding everything else constant, those who consider themselves as laggards are likely to spend .772 hours (about 46 minutes) less on their mobile device per day than the majority and/or those who consider themselves as early adopters. The next predictor is regular use of mobile gaming apps. Holding everything else constant, those who play mobile games at least once a week are likely to spend .491 hours (about 29 minutes) more on their mobile device per day than those who don't. Finally, the last significant predictor is gender. Holding everything else constant, female students are likely to spend .464 hours (about 28 minutes) more on their mobile device per day than male students.

Some findings in this study align well with the literature. By 2016, people in the United States spend most of the five hours average spent on mobile device to access social media, entertainment, and games [26]. In 2008, female students in Greece spent 3.93 hours (236 minutes) in average on their mobile phone, significantly more than male counterparts who spent 2.75 hours (165 minutes) in average [27]. In this study, age and year in college do not have significant effect toward time-spent on mobile device. One plausible reason is that the population is the same millennial generation born in the 1990s. The age factor might have a stronger effect if the comparison was made with much older students, such as graduate students who were presumably born in the 1980s and later or with much younger ones, such as high school or middle school students born in 2000s and later. The same is true for the place of origin. Since these students are still living in the same area, it would be better if the samples were taken from different geographical location, instead.

Table 2. OLS regression estimates of time spent on mobile device daily (in hours)

Gender *Female*	**.464*** **.155** **(.190)**
Attitude Toward ICT *Laggards*	**-.772*** **-.167** **(.353)**
Socioeconomic Status (SES) *Owner of the most expensive smartphone/tablet (High SES)*	**.663***** **.177** **(.186)**
Social Media *Use social media mobile apps at least once a week*	**3.010***** **.333** **(.786)**
Mobile Gaming *Play mobile games at least once a week*	**.491*** **.157** **(.182)**
Constant	1.616* - (.778)
R^2	**.279*****
Highest VIF	**1.04**
Mean VIF	**1.03**
Ramsey RESET Test	**.422**
Observation	**182**

Note: The first number reported is the unstandardized coefficient, the second number reported is the standardized coefficient and the third number reported between parentheses is the robust standard error; *p < .05. **p < .01 ***p < .001

4. Conclusion

In this paper, five significant predictors of time-spent on mobile device are found. In descending order from the strongest, they are 1) regular use of social media apps, 2) socioeconomic status, 3) negative attitude toward ICT adoption, 4) regular use of mobile gaming apps, and 5) gender. Based on the OLS regression result, females, students with the most expensive mobile device, students who use social media apps at least once a week, and students who play mobile game at least once a week are more likely to spend more time on their mobile device. Meanwhile, students who have negative attitude toward ICT adoption, or in other words those who consider themselves as laggards are less likely to spend more time on their mobile device. Furthermore, this study does not find any significant effect of age, major, year in college, place of origin, and regular use of educational apps to the time spent on mobile device.

While the findings in this study can provide a good insight into daily mobile device use among university students in Indonesia, it didn't capture a difference in the geographical location yet. Better and bigger samples from several different regions of the country might be able to overcome the limitation in this study. Also, as it turns out that homogeneity in the generation might be in play, a comparison with older and younger students or with overall Indonesian population is needed in getting the big picture of how long Indonesian people spend their time on their mobile device and what factors influence it.

References

[1] Kemp S 2014 *Social, Digital & Mobile Worldwide in 2014* [Online] Available: http://wearesocial.com/uk/special-reports/social-digital-mobile-worldwide-2014.
[2] Kemp S 2015 *Social, Digital & Mobile Worldwide in 2015* [Online] Available: http://wearesocial.com/uk/special-reports/social-digital-mobile-worldwide-2015.
[3] Kemp S 2016 *Digital in 2016* [Online] Available: http://wearesocial.com/uk/special-reports/digital-in-2016.
[4] Kemp S 2017 *Digital in 2017: Global Overview* [Online] Available: https://wearesocial.com/special-reports/digital-in-2017-global-overview [Pristup: 20. 3. 2017.].
[5] Pratama A R and Al-Shaikh M 2012 *Relation and Growth of Internet Penetration Rate with Human Development Level from 2000 to 2010* (Communications of the IBIMA).
[6] Kim S H, Mims C, and Holmes K P 2006 An introduction to current trends and benefits of mobile wireless technology use in higher education *AACE J.* **14(1)** pp 77-100.
[7] Fernández-López Á, Rodríguez-Fórtiz M J, Rodríguez-Almendros M L and Martínez-Segura M J 2013 Mobile learning technology based on iOS devices to support students with special education needs *Computers & Education* **61** pp 77-90.
[8] Hwang G J, Wu P H, and Ke H R 2011 An interactive concept map approach to supporting mobile learning activities for natural science courses *Computers & Education* **57(4)**.
[9] Martin F and Ertzberger J 2013 Here and now mobile learning: An experimental study on the use of mobile technology *Computers & Education* **68** pp 76-85.
[10] Ally M 2009 *Mobile learning: Transforming the delivery of education and training* (Athabasca University Press)
[11] Goksu I and Atici B 2013 Need for mobile learning: technologies and opportunities *Procedia-Social and Behavioural Sciences* **103** pp 685-69.
[12] Hong F Y, Chiu S I, and Huang D H 2012 A model of the relationship between psychological characteristics, mobile phone addiction and use of mobile phones by Taiwanese university female students *Computers in Human Behavior* **28(6)** pp 2152-9.
[13] Kamibeppu K and Sugiura H 2005 Impact of the mobile phone on junior high-school students' friendships in the Tokyo metropolitan area *Cyberpsychology & Behavior* **8(2)**.
[14] Ha J H, Chin B, Park D H, Ryu S H, and Yu J 2008 Characteristics of excessive cellular phone use in Korean adolescents *CyberPsychology & Behavior* **11(6)** pp 783-4.
[15] Chóliz M 2012 Mobile-phone addiction in adolescence: the test of mobile phone dependence (TMD) *Prog Health Sci.* **2(1)** pp 33-44.
[16] Ozkan M and Solmaz B 2015 Mobile Addiction of Generation Z and its Effects on their Social

Lifes:(An Application among University Students in the 18-23 Age Group) *Procedia-Social and Behavioral Sci.* **205** pp 92-8.

[17] Young K S and Rogers R C 2009 The relationship between depression and Internet addiction *CyberPsychology & Behavior* **1(1)** pp 25-8.

[18] Nalwa K and Anand A P 2004 Internet addiction in students: A cause of concern *CyberPsychology & Behavior* **6(6)** pp 653-6.

[19] Salehan M and Negahban A 2013 Social networking on smartphones: When mobile phones become addictive *Computers in Human Behavior* **29(6)** pp 2632-9.

[20] Pratama A R 2017 Cross-country Analysis of ICT and Education Indicators: An Exploratory Study *IOP Conference Series: Materials Science and Engineering* **185(1)** (IOP Publishing).

[21] Pratama A R 2017 Exploring personal computing devices ownership among university students in indonesia *Int. Conf. on Social Implications of Computers in Developing Countries* (Cham: Springer) pp. 835-41.

[22] Poushter J 2016 Smartphone ownership and internet usage continues to climb in emerging economies *Pew Research Center* **22**.

[23] Goerke V and Oliver B 2007 Australian undergraduates' use and ownership of emerging technologies: Implications and opportunities for creating engaging learning experiences for the Net Generation *Australasian J. of Educational Technol.* **23(2)** pp.171-86.

[24] Kobus M B, Rietveld P, and Van Ommeren J N 2013 Ownership versus on-campus use of mobile IT devices by university students *Computers & Education* **68** pp 29-41.

[25] Margaryan A, Littlejohn A, and Vojt G 2011 Are digital natives a myth or reality? University students' use of digital technologies *Computers & Education* **56(2)** pp 429-40.

[26] Khalaf S and Kesiraju L 2017 U.S. consumers time-spent on mobile crosses 5 hours a day *Flurry Analythics Blog* [Online] Available: *http://flurrymobile.tumblr.com/post/157921590345/us-consumers-time-spent-on-mobile-crosses-5*

[27] Economides A A and Grousopoulou A 2008 Use of mobile phones by male and female Greek students *Int. J. of Mobile Communications* **6(6)** pp 729-49.

Design of Telegram Bots for Campus Information Sharing

Hari Setiaji[1], Irving V Paputungan[2]

[1,2] Informatics Department, Universitas Islam Indonesia

[2]Informatics Department, Universitas Islam Indonesia
Universiti Teknologi Petronas, Malaysia

[1]hari.setiaji@uii.ac.id, [2]irving@uii.ac.id

Abstract. This paper presents a design of Telegram bot to support Campus information sharing. The communication method used is Webhooks. Webhooks is able to provide zero latency and handle multiple requests concurrently during communication within the Telegram bot. Such bot serves the information via specific commands. The Telegram bot prototype shows that even though Webhooks is able to provide information as requested, Webhooks setting is difficult and trickier.

Keyword: telegram, telegram bot, webhooks, telegram command

1. Introduction

Technology-enabled campuses are critical to a university's survival. Aside from the obvious reasons of the students and staff using technology on a daily basis, handheld and mobile devices have also redefined traditional higher education culture. With mobile devices in hand, any student can find the answers to almost any of life's questions within seconds.

One of the most helpful features in the mobile devices is messenger. It becomes the main tool to communicate each other, to share information and to interact. As Short Message Service (SMS) becomes old-fashioned, people normally use messenger application such as WhatsApp, Blackberry Messenger (BBM), and Telegram to perform direct communication. BBM was very popular in the late 2000s and after a decade, it started to lose out to WhatsApp. WhatsApp was incorporated in 2009 and becomes a top chat app in 109 countries. Facebook then acquired WhatsApp in 2014 to make a bigger play in the rapidly-growing messaging market, along with its own messenger platform. Recent apps called Telegram works much like the WhatsApp, using end-to-end encryption to protect shared information (Pinto, 2014) (Hamburger, 2014). The company claims to have over 100 million users, and offers features like self-destructing messages to ensure greater privacy. Users can communicate directly, either in private groups or on channels.

Latest additional feature in Telegram is Chat bots. It is a piece of program, which is based on AI and machine learning in small level. User can interact with bots by sending commands and they will reply with exactly what user needs as chatting with a knowledgeable friend. Such bots are customizable depending on user requirements and integrated into various services to, such as, controlling smart home, customized tools and other social services. In this paper, a Telegram bot to support campus information sharing is presented.

The paper is organized as follows. Section 2 presents the related work that studied telegram in the past. The case study and the system requirements are provided in the section 3. The telegram bot design can be seen in section 4. In section 5, simulation is shown. Preliminary findings based on the simulation are presented in section 6 and lastly, section 7 provides the summary of the paper.

2. Related Word

There are several previous studies on the use of Telegram. Jakobsen (2015) performed cryptanalysis on the Telegram communication. The author was able to analyse and depicts the security and encryption protocol in the Telegram. Sutikno et al. (2016) compared WhatsApp, Viber and Telegram. The result shows that Telegram offers better back-up feature and security then WhatsApp and Viber. The authors added that Telegram is the best messenger platform among those three apps. In the area e-learning, Hafiz (2015) integrates LMS e-learning and Moodle platform using MTProto open protocol. The Telegram bot, namely ELISA (E-learning integrated short announcement) will then reply to any request from students who need feedback of their work. In addition, the MTProto is able to support real-time online class discussion. One more work on Telegram bot was done by Rohim (2016). The author developed a chat bot using long polling method to share information on any offers and events from student outlet in a campus. However, with the increasing number of request that is sent to the bots, latency becomes an issue in that work. In our work, different communication method, namely Webhooks method, is used within a Telegram bot to overcome the latency issue when sharing information in campus.

3. Case Study and Requirements

Informatics Department in Islamic University of Indonesia (TF UII) takes the advantage of IT development to deliver the information to all academics staff and students. They use web portal and Facebook timeline as the main information sharing media. The main drawback is users do not frequently check such media to obtain real-time info. The proposed Telegram bot will be implemented in TF UII to provide better campus services.

The Telegram bot is required to provide information such as: TF UII profile, upcoming events, and class or mentoring timetable. The bot will broadcast information to the registered accounts regularly. Figure 1 shows the general framework of the telegram bot. A staff from TF UII will regularly share the information gathered from portals and FB page to the students, staffs, and lecturers in the campus via the Telegram bot.

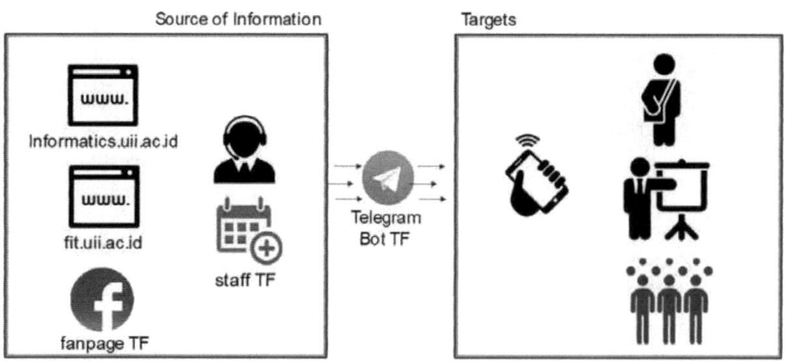

Figure 1. General Framework of Telegram bot

4. Proposed Design of the Telegram Bot

Based on the aforementioned framework, a use case is developed as shown in Figure 2. Both telegram users (student and teacher) will have to initially type the command in order to get the response from the bot. The bot obtains the information from the available source and log activity. To check if the request sender is authorized, the bot will refer to group and user management. Figure 3 illustrates how the system prototype is developed. MySQL as the database system, Python as the programming language and protection mechanism have incorporated within the bot to perform the information dissemination via Telegram.

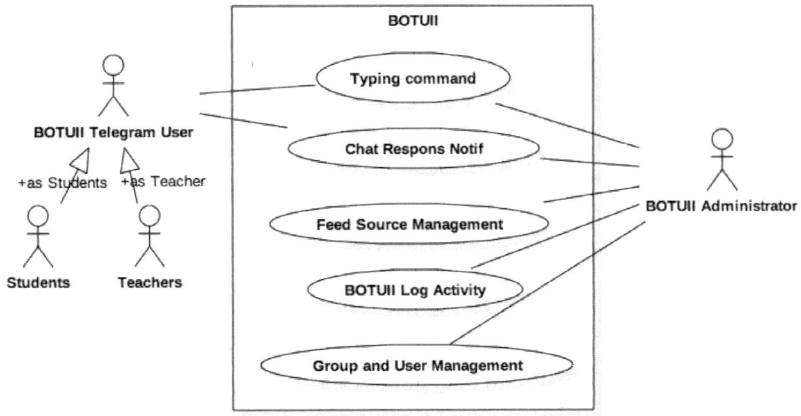

Figure 2. Bot UII use case diagram

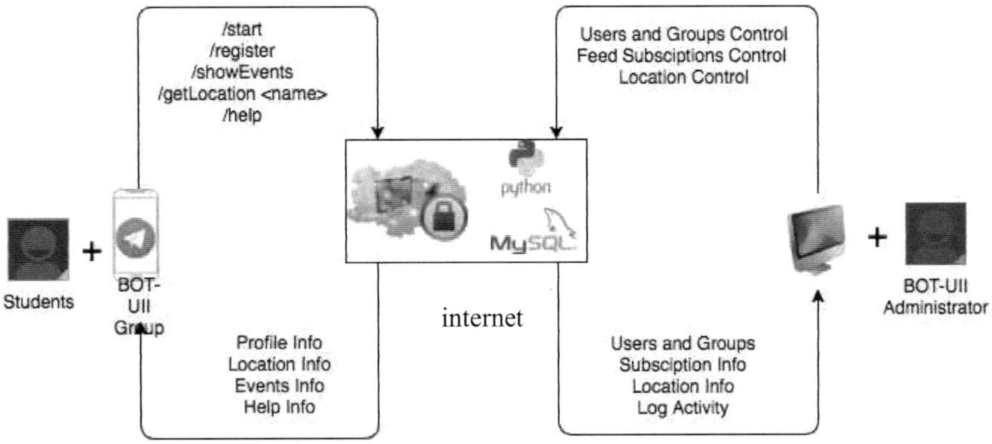

Figure 3. Telegram bot prototype

There are two (2) communication methods for communication within Telegram bot, they are Webhooks (7) and Long polling (8). Basically, Webhooks method is easier to be developed, HTTP based request and fast when responding to a request. However, it needs a dedicated server with public IP while long polling needs no dedicated server. Long polling method reduces the cost of creating home automation. It can support large number of updates without necessarily having an API call for every update. Such home automation can be more secure with additional devices. In this implementation, Webhooks is then chosen because the requested information must be fast responded and live 24/7.

IOP Conf. Series: Materials Science and Engineering **325** (2018) 012005 doi:10.1088/1757-899X/325/1/012005

5. Simulation

The telegram bot works in two ways: 1) as an information media and 2) as a position tracker. As the information media, the bot will reply any commands from users in the campus. The bot will automatically validate the information gathered from each URLs given. Figure 4 shows one simulation of command and replied message. As the position tracker, the bot can also keep the location given by the lecturer and forward the position to the students who want to meet the lecturer. Figure 5 and 6 depicts the simulation of inquiring a position using chat-bot telegram. Figure 7 shows the web-based platform for the system administrator to manage the bot.

```
Command : /showEvents

Bot: Informatika Berprestasi, waktu dan tempat dll….
(sumber: informatics.uii.ac.id)
```

Figure 4. Simulation of bot as the information media

```
Mahasiswa : /getlocation Setiaji

Chat bot : Ini lokasi terakhir dari Setiaji
https://www.google.com/maps?q=-7.743144,110.451113&ll=-7.743144,110.451113
```

Figure 5. Simulation of bot as the position tracker

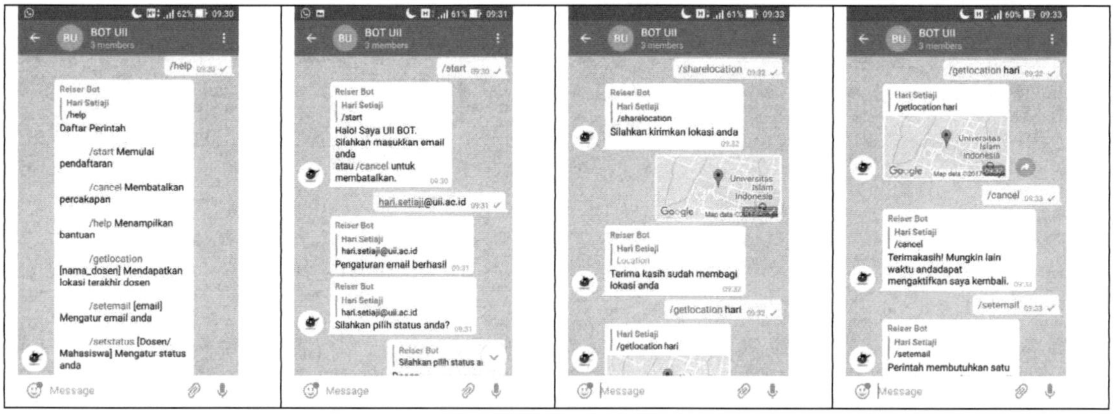

Figure 6. Full Chat bot telegram platform

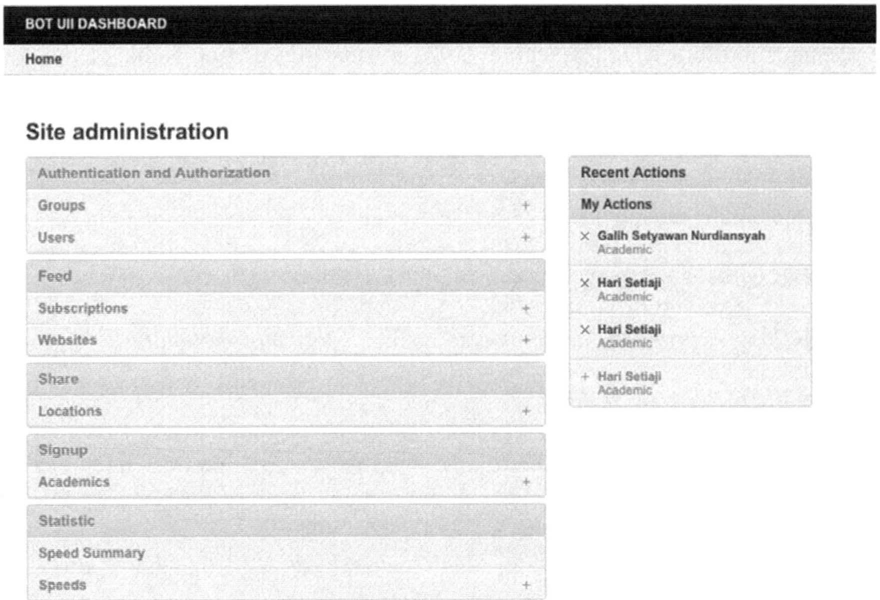

Figure 7. Web-based bot dashboard

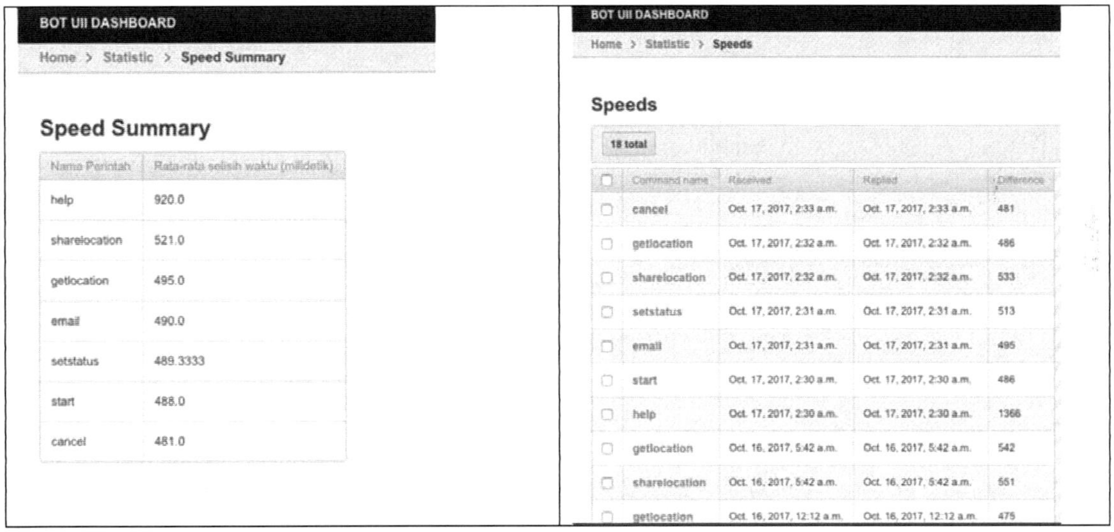

Figure 8. Response time measurement on several requests

6. Findings

There are some preliminary findings after the simulation of Telegram bot prototype. Telegram bot is able to aggregate all the information stored in the database and deliver such information to the users as requested. With some extension during code development, location can also be collected by the bot from the users and visualised clearly to the users. Analysis on the bot performance has also been conducted as shown in Figure 8. Response time was measured on different cases to see how fast the bot is responding to the requests. Bot is able to respond most of the requests in less than 0.5 seconds which is considerably satisfactory.

IOP Conf. Series: Materials Science and Engineering **325** (2018) 012005 doi:10.1088/1757-899X/325/1/012005

7. Summary

In this paper, design and prototype of Telegram bot is presented. The bot is able to perform repeated simple and complicated tasks, such as invoking and validating the information available in the database. Communication with the bot uses very simple commands and is supported by Webhooks method. Using such method, visual interface can also be developed. The development of visual interface for the chat-bot becomes the future work.

References

[1] Hafiz M 2015 ELISA: E-Learning Integrated Short Announcement *Colloquium in Computer and Math. Sci. Education* p 110.

[2] Jakobsen J B 2015 *A Practical Cryptanalysis of The Telegram Messaging Protocol* (AARHUS University).

[3] Rochim H N 2016 *Rancang Bangun Telegram Bot pada Telegram Messenger dengan Metode Long Polling untuk Koperasi Kopma UGM* (Universitas Gadjah Mada).

[4] Sutikno T, Handayani L, Stiawan D, Riyadi M A, Much I, and Subroto I 2016 WhatsApp, Viber and Telegram: which is the best for instant messaging? *Int. J. of Electrical and Computer Eng. (IJECE)* **6(3)** 909–14. http://doi.org/10.11591/ijece.v6i3.10271

[5] Pinto R 2014 *Secure Instant Messaging* Master Thesis (Department of Computer Science and Engineering, Frankfurt University).

[6] Hamburger E 2014 *Why Telegram has become the hottest messaging app in the world - The Verge* [online] Available: http://www.theverge.com/2014/2/25/5445864/telegram-messenger-hottest-app-in-the-world [Accesed 16-Feb-2017].

[7] Webhooks - https://webhooks.pbworks.com/w/page/13385124/FrontPage

[8] Long Polling - https://tools.ietf.org/html/rfc6202

Millennial Filipino Student Engagement Analyzer Using Facial Feature Classification

R Manseras[1], F Eugenio[2], and T Palaoag[3]

[1]University of the Immaculate Conception, Davao City

Isabela Colleges Incorporated, Cauayan City, Isabela

University of the Cordilleras, Baguio City

[1]roymanseras@gmail.com [2]franciscojreugenio102683@gmail.com
[3]tpalaoag@gmail.com

Abstract. Millennials has been a word of mouth of everybody and a target market of various companies nowadays. In the Philippines, they comprise one third of the total population and most of them are still in school. Having a good education system is important for this generation to prepare them for better careers. And a good education system means having quality instruction as one of the input component indicators. In a classroom environment, teachers use facial features to measure the affect state of the class. Emerging technologies like Affective Computing is one of today's trends to improve quality instruction delivery. This, together with computer vision, can be used in analyzing affect states of the students and improve quality instruction delivery. This paper proposed a system of classifying student engagement using facial features. Identifying affect state, specifically Millennial Filipino student engagement, is one of the main priorities of every educator and this directed the authors to develop a tool to assess engagement percentage. Multiple face detection framework using Face API was employed to detect as many student faces as possible to gauge current engagement percentage of the whole class. The binary classifier model using Support Vector Machine (SVM) was primarily set in the conceptual framework of this study. To achieve the most accuracy performance of this model, a comparison of SVM to two of the most widely used binary classifiers were tested. Results show that SVM bested RandomForest and Naive Bayesian algorithms in most of the experiments from the different test datasets.

1. Introduction

Millennials, also known as Generation Y, has always been a buzzword to every news articles and target market of most of the product endorsements. The term Millennials refers to a person reaching adulthood in the early 21st Century [1]. They are known to be electronic-oriented and internet-dependent people.

In the Philippines, Filipino millennials comprise one third of its population. A large number of these tech-savvy individuals are currently still in school and soon will they be movers of the economy in the future.

Education is the basic foundation of any individuals who want to survive as adults in this world. The quality of instruction delivery is an important element in achieving a better education system and this includes student engagement as its binding factor [2].

Student engagement determines a student's level of attention, curiosity and interest in the class. And for every teacher, improved engagement is part of his or her top priorities [3]. Student engagement has three (3) incorporating components: behavioral, cognitive, and emotional. A student's conduct and participation to co-curricular and extra-curricular activities is under behavioral engagement. A student's ability to understand key concepts and skills needed for learning is what is being measured under cognitive engagement. And how students react towards the people that surround them is under emotional engagement. This paper focused on student engagement and uses facial features to find out what is the current affect state of the students.

The incorporation of technology in education is among the new approaches that are being applied to prepare students for life-long learning. In the same way, the field of computing is leaning towards the implementation of computer vision-based applications. Affective computing combined with computer vision plus education, with a focus on student engagement, is a combination that is suitable for dealing with student affect states. This, along with the use of some tools (sensors, microphones, camera, etc.) [4], can contribute to improving how teachers analyze students' engagement states through the evaluation of facial features.

Machine-learning algorithms have already been used and effectively applied, with improved percentage of accuracy, in the real-time detection of facial expressions [5]. Studies have also been made that focused on detecting learning-centred states such as confusion, excitement, flow, frustrations [6], informed, inspired, persuaded, sentimental and amused [7]. There are researches that used Facial Action Coding System (FACS) in successful implementations of emotion detection through facial features. In a study by Ekman et. al, they conducted a thorough examination of the relationship of emotions through facial features [5] called Action Units (AUs). These facial features (forehead, nose, mouth, eyes, etc.) [5] were linked to certain emotions and a connection was found on some of these AUs with student engagement.

In this paper, the researchers aim to integrate affective computing and computer vision to aid teachers in gauging quality of instruction delivery by analyzing Filipino millennials' affect state, specifically student engagement during class discussion, through facial feature classification.

Teachers could best identify student engagement during class discussion. Hence, the researchers would like to prove that initial dataset labeled by teachers would be effective as basis in classification of student engagement by the proposed Predictive Classification Model.

2. Methodology

Figure 1 illustrates the methodology used by the researchers. The model served as a guide for the researchers to perform the classification and identification of the student engagement.

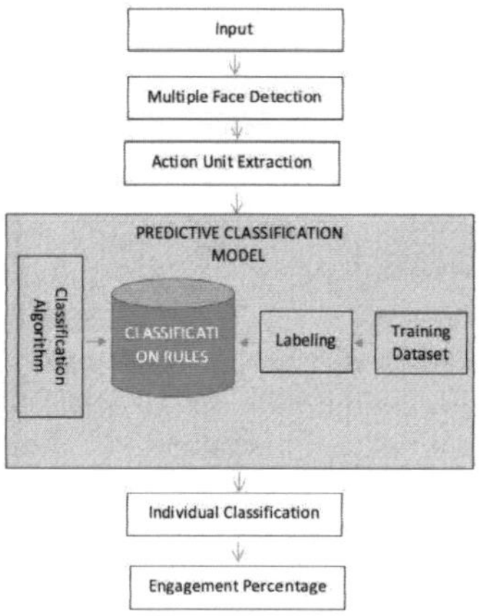

Figure 1. Conceptual Framework

2.1. Input

The input (see figure 1) would be a still image captured through a camera in a classroom environment where students are front facing the teacher during a class discussion. To address biased image capture of the class' affect state, system setting could be set to capture an image randomly between the specified time range within class discussion. Captured image will then be loaded for the multiple face detection.

2.2. Multiple Face Detection

To optimize the detection of multiple faces of students in class, an advanced machine-learning framework has been used for this experiment. The researchers used Face API which is one of the products of Microsoft Cognitive Services. The latter utilizes Microsoft Cognitive Toolkit (CNTK) as its backend. It is an open-source toolkit for deep learning algorithms. This technology employs computational network (CN) which is a framework consisting of multiple components like deep neural networks (DNNs), convolutional neural networks (CNNs) and others. It has been known that this toolkit evaluates deep learning algorithms faster than other toolkits.

The Face API of Microsoft Cognitive Services offers several features like face detection, face verification, face identification, similar face searching, and face grouping. Its face detection capability could detect multiple faces in a single still image.

2.3. Feature Extraction

After all possible faces, have been detected, the system will then apply the individual faces for identifying the various facial Action Units. These AUs will then be used as parameters for the classifier.

The system will utilize OpenFace, an open source program interface that is capable of AU recognition, facial landmarks detection, head pose estimation and gaze estimation in low hardware settings. Refer to Table 1 for the different Action Units described.

Table 1. Action units detected by OpenFace

AU	Description	AU	Description
AU1	Inner brow raiser	AU14	Dimpler
AU2	Outer brow raiser	AU15	Lip corner depressor
AU4	Brow lowerer	AU17	Chin raiser
AU5	Upper lid raiser	AU20	Lip stretched
AU6	Cheek raiser	AU23	Lip tighter
AU7	Lid tighter	AU25	Lips part
AU9	Nose Wrinkler	AU26	Jaw drop
AU10	Upper lip raiser	AU28	Lip suck
AU12	Lip corner puller	AU45	Blink

2.4. Classifier Model

To create this binary classification model, authors utilized the extracted Action Units label (true or false) and used them as parameters for classification using Support Vector Machine. In this experiment, 2 machine learning algorithms were tested and compared to prove that SVM is the most appropriate for this proposed prototype.

(a) *Support Vector Machine* (SVM): SVM has been proven to be effective in some known facial recognition systems like OpenCV and CERT [9]. It is an algorithm that classifies new dataset based on the given supervised learning. SVM yields the best hyperplane that has the largest minimum distance to the given data.

(b) *Naive Bayes*: Naive Bayesian algorithm is one of the widely-used classifier by various researchers because it is easy to build and applicable to large dataset. This algorithm is based on Bayes' theorem with independence assumptions between predictors.

(c) *Random Forest*: It is a framework consist of various methods for classification and regression and other types. This framework works by establishing a number of decision trees at training time and display the class or mean production for the individual trees. Some of its best features include high accuracy even on large dataset and it could handle multiple variables without requiring reduction on attributes in high dimensional data.

The performance of these three algorithms were estimated with cross-validation. To measure the accuracy of the trained classifier, 10-fold cross-validation (CV) was used. This process will divide the given training dataset into smaller subsets. Table 2 shows that Support Vector Machine gained well in

IOP Conf. Series: Materials Science and Engineering **325** (2018) 012006 doi:10.1088/1757-899X/325/1/012006

classifying data instance as compared to the Naive Bayes and RandomForest. These results only indicate that SVM best performs on the validation data and does not imply accuracy on unseen dataset.

Table 2. 10-fold cross validation of the training dataset

Algorithm	Precision	Recall	F-Measure
Naive Bayes	0.823	0.843	0.833
SVM	0.842	0.880	0.861
RandomForest	0.871	0.839	0.855

2.5. Data Selection

Training and Validation dataset as input to the classifier model will use the Extended Cohn-Kanade (CK+) AU-Coded facial expression database. This set of images consist of 123 subjects with 593 sequences [10].

In this prototype, the model will initially use labelled training dataset with the aid of human experts, in this case, two teachers as labellers. Data label for each instance could either be Engaged or Not Engaged.

Indicators used for label 'Engaged' student as discovered in [8], include expressions that denote concentration, gestures and excitement or presence of any of the following AUs: AU7(Lid Tighter) +AU12(Lip Corner Puller), AU5(Upper Lid Raiser), AU25(Lips Part) and AU26 (Jaw Drop); otherwise, such instance will be labelled 'Not Engaged'.

Certain procedure was applied in selecting the training dataset for the binary classification to yeild better predictive model. If the assigned label differs from the 2 labelers (e.g. labeler1 assigned Engaged and labeler2 assigned Not Engaged), the image will be discarded. This process reduced the original dataset from 593 to 499 instances.

Final test data include 3 different datasets. The first 2 datasets were captured images from 2 classes with 47 and 13 visible faces respectively. Last test dataset consists of combined instances from the first 2 datasets.

2.6. System Feedback

After going through the whole process of image capture to classification, the final system feedback to the user would be a percentage of labelled Engaged students.

2.7. Accuracy Metrics

To measure the quality of the binary classifier with respect to the test dataset, Precision, Recall and F-Measures were acquired based on true positives(TP), false positives(FP) and false negatives(FN). F-measure is the weighted average of Precision and Recall, where score reaches its best value at 1 and worst at 0. Refer to Equation 1.

$$\text{F-Measure} = 2(\text{recall} \cdot \text{precision}) / (\text{recall}+\text{precision}) \quad (1)$$

3. Results and Discussions

3.1. Face Detection and Feature Extraction

Multiple face detection using Face API yielded a considerable result. Out from the 47 subjects of the first image input, only 32 were detected (see figure 2) or 68%. And from 13 students present in the image for the second input, Face API detected 10 faces or equivalent of 77%. With an overall average of 73%. The researchers observed that the following factors are causes for non-detection of faces: (a) Non-front facing position; (b) Incomplete facial features required for face detection (e.g. eyes blocked by other objects); (c) Too far or too close from the camera causing very blurry or faded face image.

Figure 2. Screen capture from actual face detection using Face API

Moreover, facial Action Units were successfully extracted by OpenFace even with the very small image output cropped from the detected images of Face API with 64x64 image size (see figure 3).

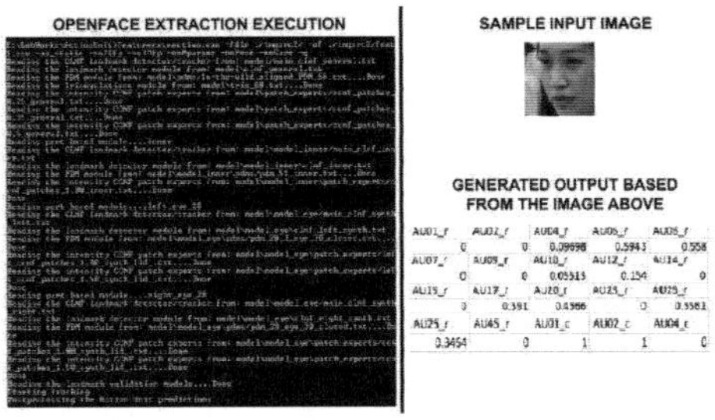

Figure 3. OpenFace Execution

3.2. Algorithm Performance with Test Datasets

The following explains the experiment results of the different algorithms, with emphasis on F-Measure, used with the final test datasets shown in tables 3,4,5.

Table 3. Accuracy of Algorithms in dataset1 with 32 instances

Algorithm	Precision	Recall	F-Measure
Naive Bayes	0.8	0.421	0.552
SVM	0.833	0.526	0.645
RandomForest	0.625	0.526	0.571

Table 4. Accuracy of Algorithms in dataset2 with 10 instances

Algorithm	Precision	Recall	F-Measure
Naive Bayes	1	0.6	0.75
SVM	0.75	0.6	0.667
RandomForest	0.4	0.4	0.4

Table 5. Accuracy of Algorithms in dataset3 with 42 instances

Algorithm	Precision	Recall	F-Measure
Naive Bayes	0.846	0.458	0.595
SVM	0.813	0.542	0.65
RandomForest	0.571	0.5	0.533

RandomForest algorithm lags behind the 2 other classifiers on all datasets with an average of 0.5 F-Measure value. Moreover, NaiveBayesian algorithm performed as the best classifier based on dataset 2 but remained on the second spot after increase in number of instances. Furthermore, SVM ranked second on the dataset2 with fewer instances but as the number of instances increased, so as its accuracy performance.

4. Conclusion and Future Works

The authors hypothesized that the use of the human experience by the teachers would best create a model based on certain indicators to detect student engagement. This assumption was proven through the F-measure achieved by the binary classifier model using the labelled training dataset. SVM achieved a very high accuracy rate based on F-measure on most of the test dataset experiments. These results imply the robustness of this algorithm as it achieved a very high precision rate in detecting correct classification in most of the instances.

Furthermore, other components of the framework need further studies to best improve this proposed prototype. In the case of multiple face detection, it is best to compare the Face API performance to other available face detection system specifically those frameworks that could address the current limitations of the CNTK. Moreover, with regards to the binary classifier model, using SVM has proven to be effective on classifying student engagement. However, authors strongly recommend that additional faces for training dataset must be sought to improve better classification and the testing of the prototype must be set to various classroom environments to assess the efficiency of the study in different test datasets.

References

[1] __ 2017 *What is Millennials (Millennial generation)? - Definition from WhatIs.com* [Online] Available: http://whatis.techtarget.com/definition/millennials-millennial-generation.

[2] Kecojevic Z 2012 Perspectives on Educational Quality: Illustrative Out- comes on Primary and Secondary Schooling in the Netherlands *CEPS J.: Center for Educational Policy Studies J.* **2(2)** p 145.

[3] Abbott S 2017 [Online] Available: http://edglossary.org/student-engagement/.

[4] Banafa A 2016 *What is Affective Computing?* [Online] Available: https://www.bbvaopenmind.com/en/what-is-affective-computing/.

[5] Al-Alwani A 2016 Estimation of Mental States Using Facial Features to Improve A Students Performance In E-Learning *Academics World52nd Int. Conf.*

[6] Grafsgaard J, Wiggins J, Boyer K, Wiebe E and Lester J 2016 Automatically Recognizing Facial Indicators of Frustration: A Learning-Centric Analysis *Humaine Association Conf. on Affective Computing and Intelligent Interaction.*

[7] McDuff D 2016 Discovering Facial Expressions for States of Amused, Per- Suaded, Informed, Sentimental and Inspired *18th ACM International Conference on Multimodal Interaction* (ACM) pp 71-5.

[8] Nyman R 2015 Indicators of Student Engagement: What Teachers Notice During Introductory Algebra Lessons *Int. J. for Math. Teaching and Learning.*

[9] Littlewort G, Whitehill J, Wu T, Fasel I, Frank M, Movellan J and Barlett M 2011 The Computer Expression Recognition Toolbox (CERT) *Automatic Face and Gesture Recognition and Workshops (FG 2011)* (IEEE) pp 298-305.

[10] Lucey P, Cohn J, Kanade T, Saragih J, Ambadar Z and Matthews I 2010 The Extended Cohn-Kanade Dataset (Ck+): A Complete Dataset for Action Unit and Emotion-Specified Expression *Computer Vision and Pattern Recognition Workshops (CVPRW)* (San Francisco, USA: IEEE) pp 94-101.

IOP Conf. Series: Materials Science and Engineering **325** (2018) 012007 doi:10.1088/1757-899X/325/1/012007

Applying Feedback Analysis on Citizen's Participation System of CSFLU Barangays on Disaster Preparedness

A J Ocampo[1], R Baro[2] and T Palaoag[3]

[1]Lorma Colleges

[2]Don Mariano Marcos Memorial State University – North La Union Campus

[3]University of the Cordilleras

ajocampo@lorma.edu, rufo.baro@gmail.com, tpalaoag@gmail.com

Abstract. Various initiatives through the use of ICT paved the way to better improve the services of the government during disaster situations. It helped in the preparation and mitigation process during disaster situations through different mediums such as Social Networking Sites and SMS to disseminate information. However, data that are gathered from this medium are not sufficient to address the problem experienced by the citizens, thus the concept of Citizen's participation system was developed. The objective of the study is to provide a mechanism or tool for barangay officials and the city government to strategically plan preventive measures during times of disasters based on the citizen's perspective, data analytics gathered from sentiments, suggestions, and feedback of the citizens was analysed using of Feedback Analysis in order to provide accuracy of data which is needed by the disaster response team that will be generated through data analytics.

1. Introduction

The Philippines is a democratic country thus active participation from its stakeholders is important. At present, promoting effective governance must involve the government unit, non-government organizations (NGO's), citizens, and private sectors to engage in collaborative activities and increase participation. (Guzman et.al, 2014). Citizen involvement plays a vital role in good governance, as the end recipient of law, policies, governance, and services; they are essential contributors to the decision-making, policy-making and service implementation given by the government. The advent of technology paved the way in bridging the gap between the government and its citizens.

Disaster situations are the most resilient times where the need for citizen involvement is required. Times of earthquake, typhoons, flooding, storm surge and other natural phenomenon are the instances where citizens are strongly affected and the need to be informed ahead of time is one of the goals of the government. Activities on disaster preparedness were implemented to local government units are one of the advocacy the government constantly plead to its citizens. However, the involvement of the community is limited because the view of the people to this advocacy is a mere implementation of the law which often leads to a reactive response to situations that results in a broader scale type of problems such as toll on death, loss of assets and greater liabilities.

According to state weather bureau Philippine Atmospheric Geophysical and Astronomical Services Administration (PAGASA), the Philippines alone is visited by at least 20 tropical cyclones every year.

Add to that the number, landslides, flood, the threat of storm surge or tsunamis and earthquakes also devastate the whole country. The chance for prevention to these disasters is very thin or worst, very remote. But the effects of these natural phenomena can be greatly reduced by educating and involving the citizenry for mitigation and adaptation.

The key to addressing such problems is Citizen Participation, thus, through the initiative of the government, the concept of E Participation or Citizen's Participation Systems was coined. In the website of E-Participation.eu (2016), the term E-Participation refers to the means of ICT-supported participation in processes concerning administration, policy making, decision making, service delivery, information provision, consultation, deliberation, etc. Through continuous ICT developments, citizens are finding new ways to participate in issues concerning them. ICT, e-government, and open government initiative paved way for e-Participation. (Tambouris et al., 2007).

Citizen's Participation System addressed 3 areas of governance during times of disaster, a.) Informative public participation which involves information dissemination on Warning signals, Class suspensions, flood status, etc.; b.) Consultative public participation which allows citizens in evaluating and assessing the performance of the barangay during times of disaster and c.) Cooperative public participation which allows the citizen to voice out their suggestions in improving the performance of the barangay (Guiao, 2016).

This study implemented Citizen's Participation system to the barangay/villages of the City of San Fernando La Union (CSFLU). It is one of the third class cities in Region 1 which is strategically located between the upland and coastal areas. It comprises of fifty-nine (59) barangays. The city always experience devastation from natural disasters, approximately thrice a year the city was visited by natural disasters such as typhoon, flooding, threat on storm surge and landslide. This study is a tool for barangay officials and the city government under the City Disaster Risk Reduction Management Council to strategically plan and implement preventive measures before and during times of disasters. The collected from citizen's feedback after analyzed provides a visual data through data analytics based from sentiments, suggestions, and feedback of the citizens. The citizen, on the other hand, will have a communications means to be informed, to collaboratively suggest and to participate in the improvement of their barangay in terms of disaster preparedness.

2. Methodology

The source of data are the citizen's and barangay profile extracted from the current Barangay Information System and the disaster preparedness policies and procedures which is based on the provisions of City Ordinance No.2011-6, Ordinance Institutionalizing the Disaster and Resiliency System in the City of San Fernando, disaster reports that the barangay submits to the City Government, and Citizen's feedback and suggestions coming from the city's official social networking site (Facebook page). Those data were gathered through series of interview to stakeholders, Survey was conducted to random citizens residing from the different barangays of the city, and Document analysis was also conducted to pattern the needed contents for report generation. This study utilized the concept of Feedback Analysis which aids analyzing the collected feedback from citizens which was generated through data analytics report. The system provides a concentrated information dissemination tool (web/ mobile accessibility) on upcoming disaster situation, Feedback tools (poll, a suggestion box or sentiment reaction box), Citizens feedback, sentiments and reactions during times of disaster (graphically represented) which aids in planning and improve the disaster preparedness and response mechanism as shown in Figure 1 below.

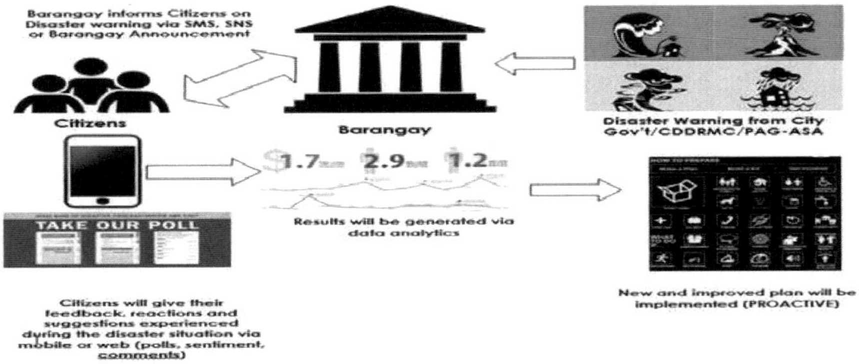

Figure 1. Conceptual Framework of Citizen's Participation System

3. Software Methodology

The development of the study was guided using Scrum Methodology, it has the capability to allow the proponents test the viability of the project through early delivery of prototypes. The incremental phases and daily team reflection allow the proponents improve the study on a daily basis. The activities are broken down into several phases, **Initiate Phase** data gathering from stakeholders was conducted, allocation of project responsibility was delegated based on prioritized product backlogs and finalization of Project timeframe was accomplished; **Plan and Estimate Phase** the project plan was drawn and gathered data were processed through feedback analysis, estimated project features were drawn and prioritized, **Implementation Phase** project development and early delivery of the prototype were done, daily meeting was conducted on the status of development; results were incremental and iterative delivery for project feedback. **Review and Retrospect Phase** project demonstration and validation were conducted for final revision. The last phase is the **Release Phase** the final delivery is done.

4. Results and Discussions

Based on interviews conducted by the proponents of this study, the city government actively participates in terms of informative public participation using Social Networking Sites specifically Facebook. Announcements on disaster-related events and situations were being posted on their city's official webpage and website. However, not all citizens are connected to this medium of communication but they still rely on TV and radio or via word of mouth on announcements especially those in the upland areas. Thus, this leads to problems on the part of the citizen such as: lack of preparation, unawareness on incoming disaster and delayed disaster response to their needs. On the part of the government: lack of information from the citizen & no knowledge on the current situation of the citizen during times of disaster leads to unimproved services during times of disaster which often results more casualties, delayed response to citizen needs and lack of trust from the citizens.

The proponents conducted interview to different citizens from Pias (upland), Poro (coastal) and Santiago (urban) the three most common disaster infested barangays. Based on the findings it was found out that these barangays are active in addressing citizens' needs during disaster situations, however, in terms of citizen participation; the citizens can only voice out their suggestions, concern or feedback during barangay/village assembly where not all the citizens attend such meetings. Also, disaster preparedness posters placed on the barangay hall is not sufficient to provide informative data on disaster preparation and mitigation because citizen's just pass through the barangay hall.

With the problems and findings stated above, the proponents applied Feedback Analysis on Citizen's Participation System in Barangays of CSFLU on Disaster Preparedness. The system allows the citizens to participate before, during and after disaster situations. It is a mechanism that aids the barangay and its officials improve their services based on the perspective, suggestions, and feedback of the citizens. This system includes polls, comment and post for the citizen's side and sentiment or reaction mechanism that can be accessed using their mobile phones. The collected data was analyzed and resulted applying

IOP Conf. Series: Materials Science and Engineering **325** (2018) 012007 doi:10.1088/1757-899X/325/1/012007

the concept of feedback analysis which would help the barangay/ village officials plan actions and strategies in improving their disaster related services specifically preparations that needs to be done before, during and after a disaster. At the same time, timely training and informing the citizens on how to prepare before an onset of a disaster affects their village could be done. Also graphical report was generated (data analytics).

4.1. Feedback Analysis Process

Feedback analysis requires the development of deep understanding of the strengths and weaknesses of an organization. This offers knowledge on how to do decision-making that would contribute to the improvement of the organization which produces results (Drucker, 2014). This form of analysis was adopted in this study to accurately assess the strengths and weaknesses of services the barangay offer before and during disaster situations. The proponents adopted the Feedback loop (see Figure 2 below) concept from the article of Imp-act (2003) which comprises of eight phases. By applying the phases of this concept on the analysis of data coming from the citizens, the barangay/ village officials, the City Disaster Risk Management Council and the city government was able to consider all the issues involved coming from the citizen's perspective and apply it to their decision making process, improvement of planning strategies and implementation of services. Thus, this results to a more responsive citizen's involvement and timely and proactive barangay official's response mechanism.

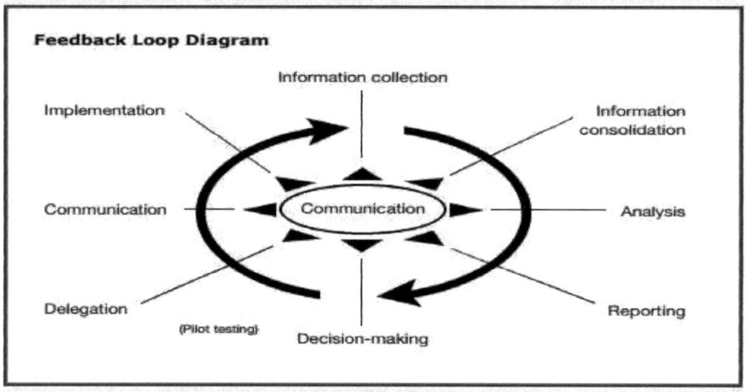

Figure 2. Feedback Loop Process

4.1.1. The result of the Feedback analysis.

The following figures below shows once the feedback analysis concept was applied to the collected data gathered from the responses of citizens.

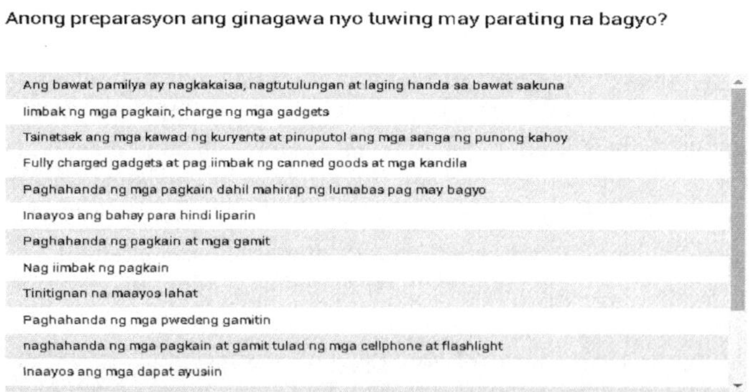
Anong preparasyon ang ginagawa nyo tuwing may parating na bagyo?

Ang bawat pamilya ay nagkakaisa, nagtutulungan at laging handa sa bawat sakuna
Iimbak ng mga pagkain, charge ng mga gadgets
Tsinetsek ang mga kawad ng kuryente at pinuputol ang mga sanga ng punong kahoy
Fully charged gadgets at pag iimbak ng canned goods at mga kandila
Paghahanda ng mga pagkain dahil mahirap ng lumabas pag may bagyo
Inaayos ang bahay para hindi liparin
Paghahanda ng pagkain at mga gamit
Nag iimbak ng pagkain
Tinitignan na maayos lahat
Paghahanda ng mga pwedeng gamitin
naghahanda ng mga pagkain at gamit tulad ng mga cellphone at flashlight
Inaayos ang mga dapat ayusiin

(What type of preparations do you make before an onset of a disaster?)

Figure 3. Citizen responses

(Does your barangay or village officials help you prepare?)

Figure 4. Result of responses (citizen response)

Figure 3. shows sample responses of citizens before and onset of a disaster. Majority of them answered that they store goods, water, charged cellular phones and power banks and prepare 1st aid kits. Figure 4. Shows the result on the response mechanism the barangay conducts during disaster preparation and majority attested that their barangay assisted them during disaster preparation.

(Is your barangay/ village active during times of disaster?)

Figure 5. Result barangay response

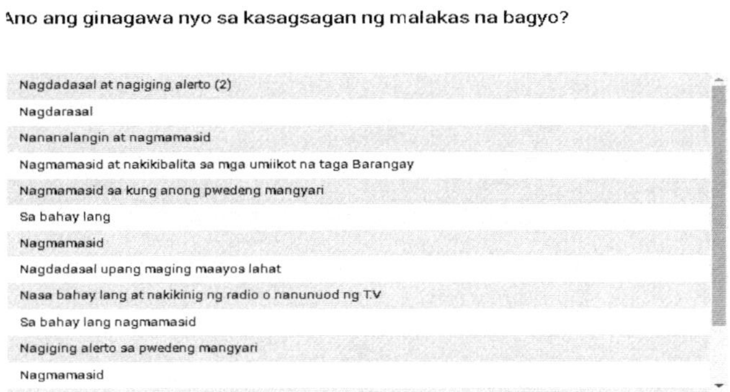

(What actions do you make during times of typhoon?)

Figure 6. Citizen responses during time of typhoon

IOP Conf. Series: Materials Science and Engineering **325** (2018) 012007 doi:10.1088/1757-899X/325/1/012007

Figure 5 shows the responses of the citizens' in terms of the village officials involvement during times of disaster and majority attested that the village officials are active in addressing their problem and needs. While Figure 6 shows the citizens' response on actions they make during times of typhoon and majority said that they are in alert state, some are praying while others are in a standby mode waiting for officials to respond to them.

4.2. Design and Features of the Citizen's Participation System

In order to address the problem the following features were incorporated the Citizen's Participation are divided into 3 areas: the Citizen's module, Barangay module, and City Government module.

4.2.1. Citizen's Module.

Though the use of the Citizen's Participation System the citizen can a.) Receive announcement and warning, b.) Post Feedback, suggestion, rate and react, c.) Collaborate with other citizens in the barangay. See figure 7– 8 for sample screenshot of the citizen's module.

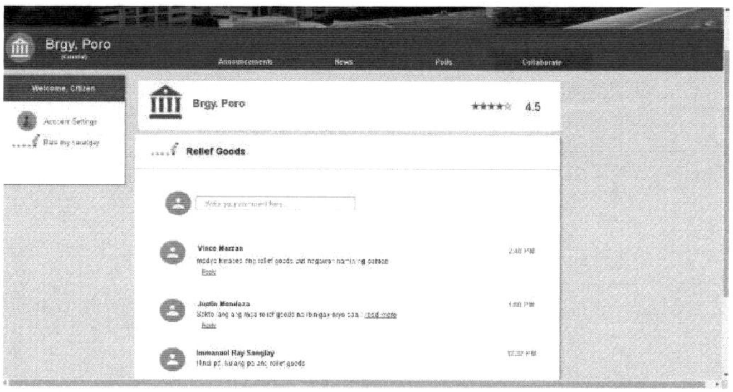

Figure 7. Citizen's Receive Message Form

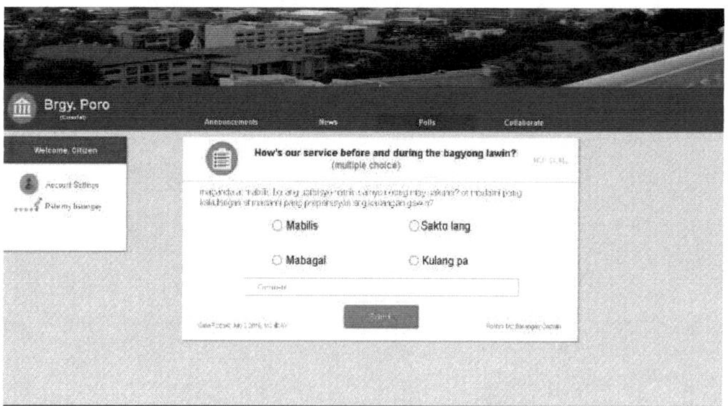

Figure 8. Citizen's Poll Module

4.2.2. Barangay Module.

Though the use of the Citizen's Participation System the barangay officials can a.) Disseminate announcement and warning, b.) Respond to citizen's concern in terms of disaster, c.) Collaborate with the citizens for improvement of services in terms of disaster preparedness, d.) Generate reports through Feedback analysed data with a data analytics output. See Figure 9 and 10.

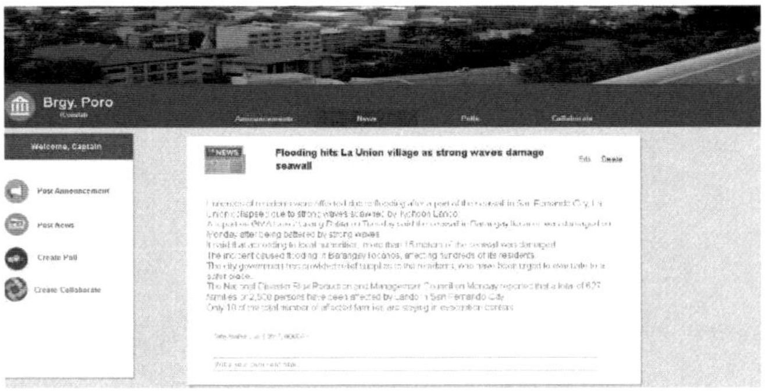

Figure 9. Barangay Information module

Figure 10. Barangay Dashboard module

4.2.3. City Government Module.
Though the use of the Citizen's Participation System the barangay officials can a.) Disseminate announcement and warning, b.) Respond to citizen's concern in terms of disaster, c.) View reports based on data analytics, d.) Monitor and supervise the barangay in terms of disaster preparation and mitigation.

5. Conclusion

The utilization of the Citizen's Participation System of CSFLU Barangays on Disaster Preparedness increased the citizen's participation in disaster preparedness, mitigation, and adaptation. Citizens are involved in the policy and decision-making through their feedbacks, reactions, and suggestions experienced during the disaster situation via mobile app and web portal. It is expected that a strong partnership will be forged between the City Government and the Barangay folks towards a zero casualty or reduced disaster risk or effect.

References

[1] Austria D, Guzman J, Siy J, and Zaplan R 2014 E-participation: bridging the gap for better governance *The DLSU Research Congress 2014* (Manila, Philippines: De La Salle University).
[2] Drucker P 2014 *The Peter Drucker Guide to Excellence.*
[3] Guiao J 2017 Human-computer interaction design for e-participation systems *Research and Natural Language Processing Workshop January 11-12.*
[4] Im-pact 2003 The Feedback Loop Responding to Clients Needs, Improving the Impact of Microfinance on Poverty *Action Research Programme* ISSN: 1740-4711.
[5] Salas J 2015 *E-Participation for Good Governance.Journal of E-Governance.*
[6] Tambouris E, Macintosh A, Aichholzer G, Winkler R, Coleman S, Wimmer M, et al *(n.d.).Introducing eParticipation* [PDF Document] Austrian Academy of Sciences - Academia.edu.

IOP Conf. Series: Materials Science and Engineering **325** (2018) 012008 doi:10.1088/1757-899X/325/1/012008

Face Detection Technique as Interactive Audio/Video Controller for a Mother-Tongue-Based Instructional Material

Excel Philip B. Guidang, Christopher John R. Llanda, Thelma D. Palaoag

Lagangilang, Abra, Philippines, Lagangilang, Abra, Philippines, Gov. Pack Rd. Baguio City, Phillippines

rieann12@gmail.com, john_llanda@yahoo.com, tpalaoag@gmail.com

Abstract. Face Detection Technique as a strategy in controlling a multimedia instructional material was implemented in this study. Specifically, it achieved the following objectives: 1) developed a face detection application that controls an embedded mother-tongue-based instructional material for face-recognition configuration using Python; 2) determined the perceptions of the students using the Mutt Susan's student app review rubric. The study concludes that face detection technique is effective in controlling an electronic instructional material. It can be used to change the method of interaction of the student with an instructional material. 90% of the students perceived the application to be a great app and 10% rated the application to be good.

1. Introduction

Face Detection is the activity of automatically detecting and locating human faces in visual media or within and image, and returns face bounding boxes. It is the first-step in face recognition systems, with the aim of extracting the face region from the background [1]. Moreover, face detection algorithms targets on the discovery of frontal human faces. Face detection can be done in several ways such as scale-friendly deep convolutional networks, robust real-time face detection, neural network-based face detection, fast cascades, tracking-learning detection, subcortical face processing, component-based face detection, Eigenfaces, and Viola-Jones algorithm [2]. Among all face detection techniques, Viola-Jones algorithm if efficient and robust for detecting frontal faces [3] which was considered in this study. The process of detecting facial features can be utilized and applied in an e-learning environment.

E-learning refers to the teaching and learning process with the use of computerize medium to facilitate classroom based instructions. With the help of modern electronic devices nowadays ranging from small to large scale computers, teachers take advantage these technology to develop instructional material to be integrated in classroom instructions[4].

Modern instructions comes with blended learning with the integration of computer vision applications to interact students' learning[4]. Face detection can be used in instructional materials especially on multimedia content topics to promote concentration and gives focus to learners. To further enhance the learning of the student, a multi-lingual capability of the multimedia was also considered in this study.

In Abra, where people have multiple dialects, tribe dialect also known as mother tongue language is used by people as the first language followed by Iloco. It is the commonly used dialect of people in Abra to communicate with other tribes. Filipino and English language is also used as a medium of instruction in the classroom. The problem is, students find lessons in programming difficult to understand when it

IOP Conf. Series: Materials Science and Engineering **325** (2018) 012008 doi:10.1088/1757-899X/325/1/012008

is delivered in English. So, language barrier is obvious. The need to develop an instructional material in mother-tongue is needed to eliminate language barrier.

Department of Education (DepEd) Memorandum No. 32, s. 2012, K-12 program obligates the use of Mother Tongue- Based Multilingual Education (MTB-MLE) in the delivery of lessons. The Abra State Institute of Sciences and Technology as partner of Dep-ED in the delivery of the K-12 Senior High School program offers TechVoc-ICT track on computer programming and Hardware servicing strand, finds the need to develop an instructional material adhering to the memorandum as it experienced language barrier.

This study proves the possibility of replacing the method of interaction from keyboard and mouse inputs which are commonly used in e-learning by a face detection Face Detection Technique. The core idea of the application is to play the embedded instructional material and when it detects a face, suspend the same when it cannot, and resumes again when it re-detects a face.This study similar from that of Ugurlu, Y., (2012) in his study Human Interactive E-learning Systems wherein Head Posture Images was used as an input.

1.1 Objectives of the Study

The researchers present how face detection technique can be used in an Interactive Audio/Video Controller in an instructional material. Specifically, it achieved the following objectives:

1. Developed a face detection application that controls an embedded mother-tongue-based instructional material for face-recognition configuration using Python;
2. Determined the perceptions of the students and faculty members using the Mutt Susan's student app review rubric.

2. Methodology

2.1. Research Design

The researchers used both descriptive and applied methods of research in the study. Descriptive research was used to narrate how the face detection application with embedded mother-tongue-based instructional material was developed and the perceptions of the students regarding the application. The study used Rapid Application Development (RAD) model to produce the Face Detection Technique as audio/video controller for an Instructional Material.

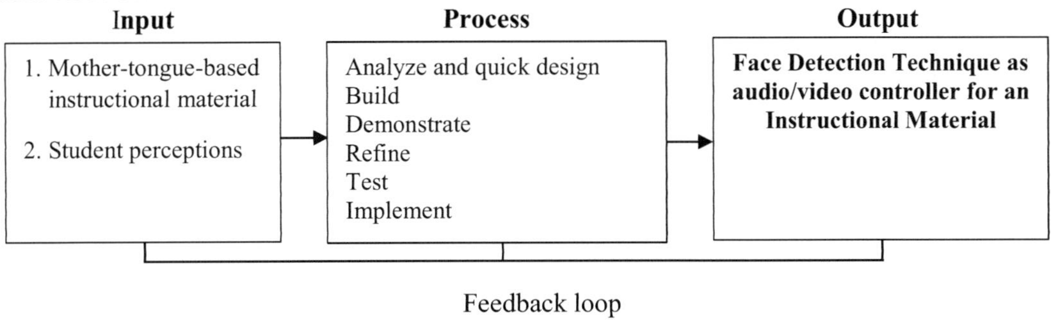

Figure 1. Paradigm of the Study

The paradigm in Figure 1 shows the interaction of the Input-Process-Output (IPO) variables. The input variables were mother-tongue-based instructional material and Student perceptions.

The Face Detection Technique as audio/video controller for an Instructional Material was processed by the following procedures: a) analyze and quick design; b) build; c) demonstrate; d) refine; e) test; and f) implementation.

The output is the The Face Detection Technique as audio/video controller for an Instructional Material . The feedback loop of the study was based from the results of the demonstration being tested and evaluated by the users or respondents which provided room for improvement.

IOP Conf. Series: Materials Science and Engineering **325** (2018) 012008 doi:10.1088/1757-899X/325/1/012008

2.2. Sources of Data

Table 1. Distribution of Respondents

Position	N
BSIT 4 Students	26
Faculty Members	4
Total	**30**

Legend: N – total population

Table 1 shows the distribution of the respondents. Twenty six (26) BSIT students and four BSIT faculty members of the Abra State Institute of Sciences and Technology served as the respondents of the study.

The respondents were asked to evaluate the face detection application with embedded mother-tongue-based instructional material.

2.3. Data Instrumentation

For objective 1, the application was developed using the Rapid Application Development model. It has six phases, namely, a) analysis and quick design; b) build; c) demonstrate; d) refine, e) testing; and f) implementation.

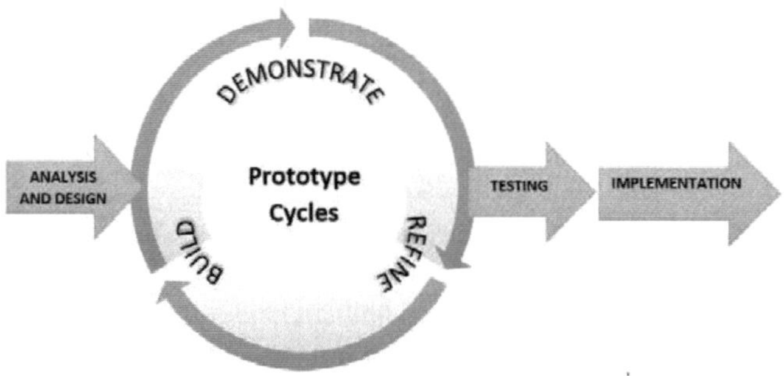

Figure 2. Rapid application Model

2.3.1. Analyze and quick design

Existing delivery method and platform of the Instructional material (IM) was examined. Moreover, viewing width of the devices used to access the School website was also considered.

The playing of the IM depends solely on the frontal face detection. The IM stops when frontal-face detection is not present and resume immediately soon after face detection.

2.3.2. Build

The mother-tongue-based IM was recorded using CamStudio and the face detection was developed using tracking.js framework. Later on, the IM was embedded in the face tracking framework.

2.3.3. Demonstrate

The application was shown to the some of the students while the application was being created to generate comments and suggestions for the improvement of the prototype.

2.3.4. Refine

As comments and suggestions comes in, these are integrated in the design of the application.

2.3.5. Testing

The application was tested in a localhost environment. Simultaneous access to the server using mobile devices such as smartphones, tablets, and laptops was conducted.

IOP Conf. Series: Materials Science and Engineering **325** (2018) 012008 doi:10.1088/1757-899X/325/1/012008

2.3.6. Implementation
The face detection app was embedded in an emulated ASIST website environment. Next, the randomly selected IT students were asked to access and use the application using smartphones, tablets, and laptops and were asked to evaluate the same later on.

For objective 2, a survey rubric designed by Mutt Susan (2014) to evaluate student applications was adopted. The Student App Review Rubric, features five sections (or criteria) that students can grade when assessing an app. Each of these criteria can be graded with a numerical number from 0 to 4 with 4 as the top grade. According to Jaafar, 2008, usability evaluation provide answers regarding whether the system was designed and developed according to the user's requirements

2.4. Data Analysis
In relation to the student and faculty perception of the application, the data gathered were interpreted using the suggested interpretation of the data like a great app is one that gets a score ranging between 20-17, a good one between 16-13, an average 12-10, Needs work 9-6, and a bad app get a score of 5-0.

3. Findings

3.1. Viola-Jones Algorithm
Several algorithms have been used and tested for computer vision on face detection technique, however, the Viola-Jones algorithm for detecting object such as face and has high accuracy rate with 98.38% using real-time video camera[8]. Figure 3 illustrates how Viola-Jones algorithm works in this study.

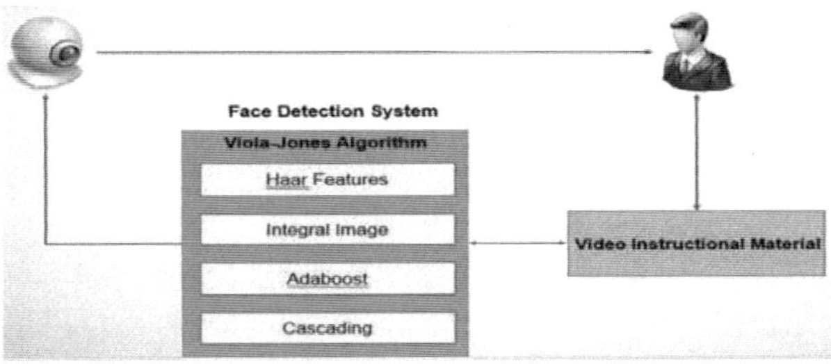

Figure 3. Viola-Jones algorithm

The initial step of Viola-Jones algorithm is the Haar-like features that has weak classifiers. Haar-like features are a simple method or tool to describe a certain object within an image. Each Haar-like feature describes the object by describing the difference in its intensities at a certain region. Therefore it can confirm whether an image contains this object by checking if certain regions have differences in their intensities similar to the object to be detected [9]. An example that would clarify this concept is the Haar like feature describing the region of eyes and cheeks within a human face. This is based on the fact that the eye and eyebrow region have darker intensities than the cheek region. Therefore if the image at this region happens to have a large difference in its intensities, this implies that the image may contain a face.

After finding all the possible Haar-like features that may exist within a window of a given size, each of these features is treated as a weak classifier and is passed on to the AdaBoost algorithm to start training. AdaBoost will also receive training images as an input and will run each given weak classifier on each input image. According to the results the weight of each classifier will be updated, and the algorithm iterates again.

A function to calculate the sums of each rectangle is called integral image. The integral image is a matrix that contains the sums of all pixels on the upper left region to any pixel. This function was tested by generating an image with the darkest level of gray (value =1), and then calculating the integral image. The output Integral Image increases in brightness as it moves towards the right and moves downwards

IOP Conf. Series: Materials Science and Engineering **325** (2018) 012008 doi:10.1088/1757-899X/325/1/012008

since the integral image sums up all the values on the upper left of the current pixel. The final step is a cascading classifiers to reduce the amount of computation and therefore finalize the result of giving a true face or not [10].

3.2. Embedding mother-tongue-based instructional material in image-recognition configuration in Python.

Now, the researchers will present the functions and considerations in the development of the application. First, the general function that manipulates how the face detection behaves relies on the following code snippet based on the tracking.js framework.

```
myTracker.on('track', function(event) {
        if (event.data.length === 0)
        {
           // No targets were detected in this frame.
        } else {
              event.data.forEach(function(data) {
                // Plots the detected targets here.
        });
        }
});
```

In the proposed face detection application, once that the camera detects a face, it activates a line x.play() function and do the x.pause() function when it cannot. Second, the coding of the face detection was ensured that all viewing devices can be used to access the E-learning material by setting the initial-scale equal to 1. Third, the important libraries that must be included in the face detection application such as tracking.min.js, face.min.js, dat.gui.min.js, state.min.js, bootstrap.mins.js, jquery.min.js, and bootstrap.min.js. These libraries ensures that tracking.js primarily works well.

In the succeeding discussion, we'll present how student will interact with the application.

Figure 4. Face Detection trigger button

Figure 4 shows the trigger button embedded in the emulated ASIST website that allows face detection using tracking.js as indicated by a blue down arrow.

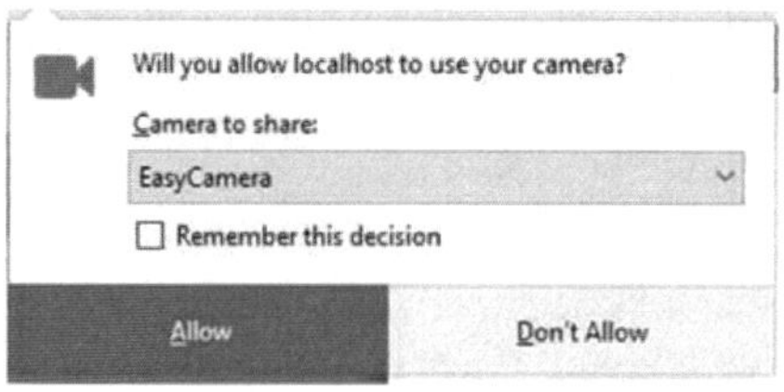

Figure 5. User approval

It can be seen on Figure 5 a dialog box that requires user approval on the use its device camera by the application.

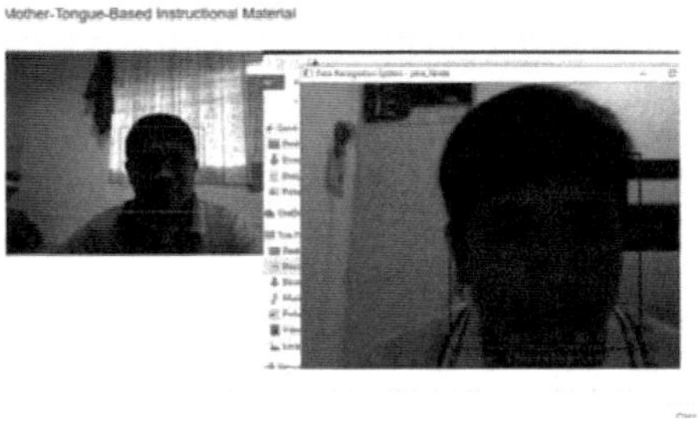

Figure 6. Face detection playing the IM

It can be seen on Figure 6 Face detection playing an IM. The application automatically plays the instructional material once it has detected a frontal-face and pause the same material when it cannot as seen on Figure 7. So, student focus on the material is a requirement in playing the instructional material. This finding corroborates the study of Ugurlu, Y., (2012) that head posture images provides adequate information in increasing human-computer interaction in e-learning system.

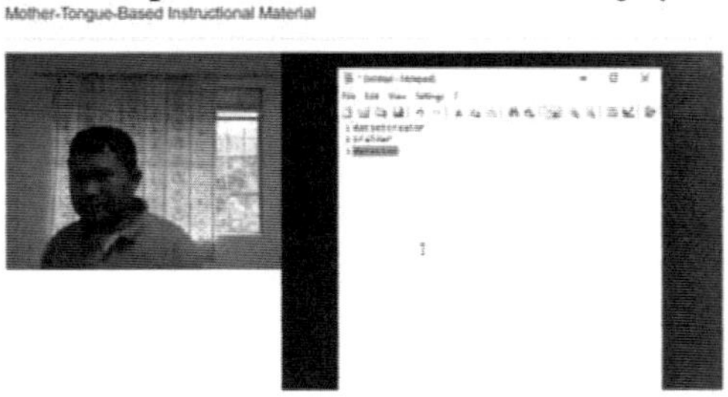

Figure 7. Face detection pausing the IM

3.3. The perceptions of the students and faculty members using the student app review rubric.

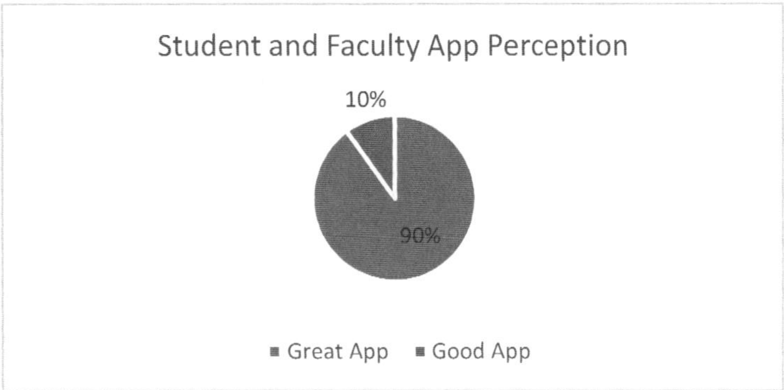

Figure 8. Student and Faculty App Perception

It can be can be seen on figure 8 on the student and faculty perceptions. 90% of the students and faculty members perceived the application to be a great app and 20% rated the application to be good. This implies that the students and faculty members agree that the face detection application is great application in controlling an instructional material.

4. Conclusion
Face detection technique can be applied in a multimedia instructional material to add excitement and it can be used to change the method of interaction of the student with an instructional material. Face detection application is great application in controlling an instructional material.

5. Recommendations
The researchers strongly recommend the improvement of this research along the areas of illumination control and face recognition in lieu of face detection. In this case, the learning is delivered to the intended learner before the frontal camera.

References

[1] Barnouti N H, Al-Dabbagn S S M, Matti W E, and Naser M A S 2016. *Face Detection and Recognition Using Viola-Jones with PCA-LDA and Square Euclidean Distance.*

[2] Nielsen J B 2010 Face detection and recognition in video-streams *Kongens Lyngby 2010* (IMM-B.Sc-2010-14).

[3] Al-Bardi S, and Al-Sawadi N 2014. *Face Detection and Recognition for Automating Class Attendance Systems* (College of Engineering Electrical and Computer Engineering Department).

[4] Zhang D, Zhou L, Bringgs R, and Nunamaker J F 2005 *Instructional Video in E-learning: Assessing the impact of interactive video on learning effectiveness.*

[5] Ugurlu Y 2013 Human interactive e-learning systems using head posture images *World Academy of Sci., Eng. and Technol.* **65** p 201.

[6] Susan M 2014 *Digital Divide and Conquer, A Great Student Rubric for Reviewing Apps*, Available: http://www.educatorstechnology.com [Accessed: 16 Sept. 2017].

[7] Jaafar A 2008 Malaysian smart school courseware usability study: the effectiveness of analytical evaluation technique compared to empirical study *WSEAS Transactions on Information Science and Applications* **5(4)** pp 342-8.

[8] Chintalapati S and Raghunadh M V 2013 *Automated Attendance Management System Based On Face Recognition Algorithms.*

[9] Viola P and Jones M J 2004 Robust real-time face detection *Int. J. of Computer Vision* **57(2)** pp 137–54,

[10] Makovetsky R and Petrosyan G 2013 Face detection and tracking with web camera *Fundamentals of Computer Vision.*

Indexing the Environmental Quality Performance Based on A Fuzzy Inference Approach

Lizda Iswari

Department of Informatics, Universitas Islam Indonesia, Jalan Kaliurang Km. 14 Ngaglik Sleman Yogyakarta Indonesia

Lizda.iswari@uii.ac.id

Abstract. Environmental performance strongly deals with the quality of human life. In Indonesia, this performance is quantified through Environmental Quality Index (EQI) which consists of three indicators, i.e. river quality index, air quality index, and coverage of land cover. The current of this instrument data processing was done by averaging and weighting each index to represent the EQI at the provincial level. However, we found EQI interpretations that may contain some uncertainties and have a range of circumstances possibly less appropriate if processed under a common statistical approach. In this research, we aim to manage the indicators of EQI with a more intuitive computation technique and make some inferences related to the environmental performance in 33 provinces in Indonesia. Research was conducted in three stages of Mamdani Fuzzy Inference System (MAFIS), i.e. fuzzification, data inference, and defuzzification. Data input consists of 10 environmental parameters and the output is an index of Environmental Quality Performance (EQP). Research was applied to the environmental condition data set in 2015 and quantified the results into the scale of 0 to 100, i.e. 10 provinces at good performance with the EQP above 80 dominated by provinces in eastern part of Indonesia, 22 provinces with the EQP between 80 to 50, and one province in Java Island with the EQP below 20. This research shows that environmental quality performance can be quantified without eliminating the natures of the data set and simultaneously is able to show the environment behavior along with its spatial pattern distribution.

Keywords: Mamdani FIS, Land Cover, River Index, Water Index, Spatial

1. Introduction

Environmental degradation can be stated as a disaster that causes losses for the community, not merely in the economic aspects but a greater impact may also attach to the human lives. Efforts to reduce the rate of environmental degradation and restoration continue by both the government and community organizations.

In some countries, the environmental quality performance are monitored and assessed in many aspects, such as environmental dimension in quality of life [1], environmental awareness and behavior in terms of the social-psychological factors [2], and environmental quality health index for cities based on the noise levels and the air pollutant concentration [3].

Starting in 2009, the Indonesia Ministry of Environment has also developed a similar monitoring instrument named Environmental Quality Index (EQI) [4]. This instrument consists of three key indicators, namely river water quality, air quality, and land cover/vegetation. Currently, the determination of EQI uses a simple weighting formula in which the water and air quality are weighted 30% respectively, and the land cover index holds 40% to represent the environmental quality performance at the provincial level [5].

Data of environmental quality index is closely related to the continuous nature of phenomenon on the earth surface. The interpretation of environmental conditions may contain some uncertainties and has a range of circumstances that possibly is partially true, i.e. data of environmental indicators that may be interpreted as "between" poor and good classification, and hence will be less appropriate if processed under a common statistical approach. As mentioned in [6], a system that has a certain degree of complexity will be impossible to provide an accurate and meaningful description of system behavior by using a quantitative method. Moreover, we analyzed that the current of EQI determination has not provided some inferences yet to inform dominant factors that lead a province to have a good or poor environmental quality performance.

One approach to overcome these problems can be done by applying the Fuzzy Inference System (FIS). It has a similar way of reasoning such as human thinking and can be effectively used to describe approximate and uncertain phenomena in the real world [7]. The FIS outcome may provide outstanding results, especially when the existing system cannot be interpreted by quantitative methods due to its complexity nor has inaccurate and uncertain qualitative information [6]. Research with regard to FIS can be found in [6] diagnosing the degradation of feed water heater and in [8] determining the quality of dates.

In this research, we aim to manage the indicators of EQI with a more intuitive computation technique that is able to handle an object with fuzzy natures and makes an inference related to the environmental quality performance. Research was conducted in three stages based on the Mamdani Fuzzy Inference System (MAFIS) approach, i.e. fuzzification, data inferencing, and defuzzification. The selection on Mamdani is based on its ability to receive human input as well as its advantages on synthesizing a set of linguistic control rules which can be obtained from experienced human operators [8].

2. Environmental Parameters

The EQI instrument used in Indonesia consists of three indicators and specified into 10 parameters [5][4].

2.1. The River Quality Parameters

The river index commonly quantified through these seven parameters as follow:

a. Total Suspended Solid (TSS): a residue of total solids retained by a filter and working to estimate the rate of sedimentation. The threshold value of good TSS is 50 mg/L.
b. Dissolved Oxygen (DO): the amount of oxygen contained in the water. Water that has a high DO showed a low level of contamination with the threshold value of 4 mg/L.
c. Chemical Oxygen Demand (COD): the amount of organic compounds in the water equivalent to the amount of oxygen to oxidize organic compounds chemically with the threshold value of 25 mg/L.
d. Biochemical Oxygen Demand (BOD): the amount of oxygen needed by microorganisms to decompose organic compounds under the aerobic conditions with the threshold value of 3 mg/L.
e. Total-Phosphate (TP): the amount of organic compounds, such as proteins, urea, and the result of the decomposition process. The higher TP leads to the decline in Dissolved Oxygen values with the threshold value of 0.2 mg/L.
f. Fecal Coli (FC): a group of bacteria found in domestic waste and consisting of Coliform, Eschericia coli, and Streptococcus faecalis bacteria. The threshold value was determined based on the data average in the last 3 years, i.e. 5000/100mL.
g. Total Coli (TC): the contamination caused by the human feces. Similar to Fecal Coli, the threshold value was also based on the data average in the last 3 years, i.e. 3000/100mL.

2.2. The Air Quality Parameters

The most common of air pollution mainly come from the combustion process which includes fossil fuels, such as Nitrogen Oxides (NO), Carbon Monixide (CO), and Sulfur Dioxide (SO). For indicating the air quality performance, the Ministry of Environment has set NO2 and SO2 as the parameter. The threshold value of the healthy air for NO2 should not exceed 100μg/m3 and SO2 is 60μg/m3.

2.3. The Land Cover

The land cover is the condition of the earth's surface that describes the appearance of land cover and vegetation. The state of land cover in Indonesia has been obtained as the interpretation of satellite images Landsat 8 OLI. This interpretation is divided into two major classifications, i.e. the forested areas and non-forested areas. In this research, we took the total forested areas as the indicator of the land cover with the threshold value of 40%.

3. Fuzzy Inference System

Fuzzy refers to an adjective meaning of vague or unclear [9]. The fuzzy logic is introduced as a form of basic mathematical logic development which provides a methodology to model uncertainty and inaccurate qualitative information [6]. To develop a system with fuzzy logic model, [9] described three core elements of fuzzy inference system.

3.1. Fuzzification

Fuzzification deals with a set of process to alter crisp inputs into fuzzy values using the membership function. To fuzzify the data input, we have to define the fuzzy variables along with its fuzzy sets. Fuzzy set is a description of the common condition of fuzzy variable.

We applied 10 parameters of EQI as the fuzzy input variables, i.e. Total Suspended Solids (TSS), Dissolved Oxygen (DO), Chemical Oxygen Demand (COD), Biochemical Oxygen Demand (BOD), Total Phosphate (TP), Fecal Coli (FC), Total Coli (TC) belong to the river index parameters, NO2 and SO2 belong to the air index parameters and the coverage of land cover (LC). On each of river index and land cover, we defined two types of fuzzy sets, i.e. Good and Poor. Meanwhile for the air index parameters it had Good and Medium fuzzy sets. For the fuzzy output variable, we used the terminology of Environmental Quality Performance (EQP) and defined by having 3 types of fuzzy sets, i.e. High, Moderate, and Poor.

Afterward we defined fuzzy set domain as the allowed value to be processed in a fuzzy set. We considered the threshold and standard deviation of the data set to define data domain of input variables. In these, the threshold value worked as the center value and the standard deviation worked as the range of each fuzzy set. For instance, on TSS we found that the threshold value was 50 mg/L and the standard deviation was 36.5 mg/L (rounded to 40). Hence, the data domain of TSS was in the range of 30 to 70 with 50 on the center as can be seen in Figure 1. Meanwhile, the data domain of output variable was determined manually to the interval of 0 to 100.

The last step on fuzzification was to defining a fuzzy set membership function which mapped an input to a membership degree and was represented in the real number with the interval of 0 to 1. The membership function can be illustrated as curves to describe the mapping mechanism of continuous or linear data. In this research, we found all fuzzy input variables had the continuous characteristics and hence presented as the sigmoid curves. For the fuzzy output it was defined as the linear objects and was presented using the trapezoidal curves.

In more detail, for the membership function of fuzzy input we applied the growth and depreciate of sigmoid function as illustrated in Fig. (1) with the mathematical functions that are described in Eq. (1) and (2). Meanwhile, for the fuzzy output it was defined as the trapezoidal linear increase and decrease and the triangular function with the mathematical function in Eq. (3), (4) and (5).

For example, the TSS value on a province was 47.03 mg/L. By mapping this value to its membership function as Eq. (1) and (2), we found that the membership degree of TSS on Good at 0.6374 and Poor at 0.3625 as seen in Fig. (3), indicating that TSS of this province was closer to Good rather than Poor condition.

$$Growth[x;a,b,c] = \begin{cases} 0 \; for \; x < a \\ 2\left(\frac{x-a}{c-a}\right)^2 \; for \; a \leq x \leq b \\ 1-2\left(\frac{c-x}{c-a}\right)^2 \; for \; b < x \leq c \\ 0 \; for \; x > c \end{cases} \quad (1)$$

$$Depreciate[x;a,b,c] = \begin{cases} 1 \; for \; x < a \\ 1-2\left(\frac{x-a}{c-a}\right)^2 \; for \; a \leq x < b \\ 2\left(\frac{c-x}{c-a}\right)^2 \; for \; b \leq x \leq c \\ 0 \; for \; x > c \end{cases} \quad (2)$$

$$TrapLinUp[x;a,b,c] = \begin{cases} 0 \; for \; x \leq b \\ \frac{x-b}{c-b} for \; b < x < c \\ 1 \; for \; x \geq c \end{cases} \quad (3)$$

$$TrapLinDown[x;a,b,c] = \begin{cases} 1 \; for \; x \leq a \\ \frac{b-x}{b-a} \; for \; a < x < b \\ 0 \; for \; x \geq b \end{cases} \quad (4)$$

$$Triangle[x;a,b,c] = \begin{cases} 0 \; for \; x \leq a \; or \; x \geq c \\ \frac{x-a}{b-a} \; for \; a < x < b \\ \frac{c-x}{c-b} \; for \; b < x < c \\ 1 \; for \; x = b \end{cases} \quad (5)$$

where:
a = bottom boundary;
b = center point;
c = upper boundary

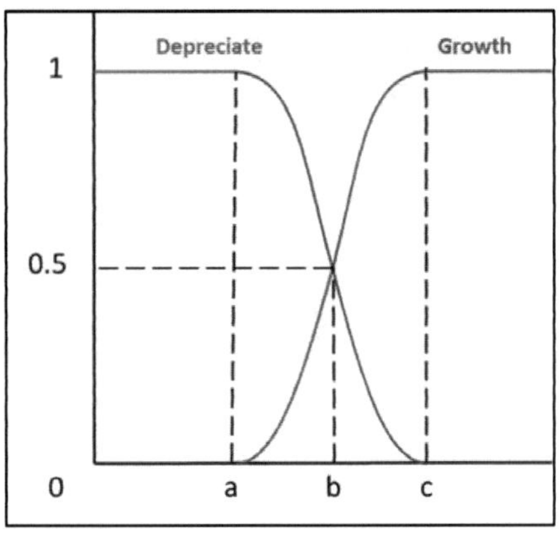

Figure 1. The Membership Curves of Fuzzy Input

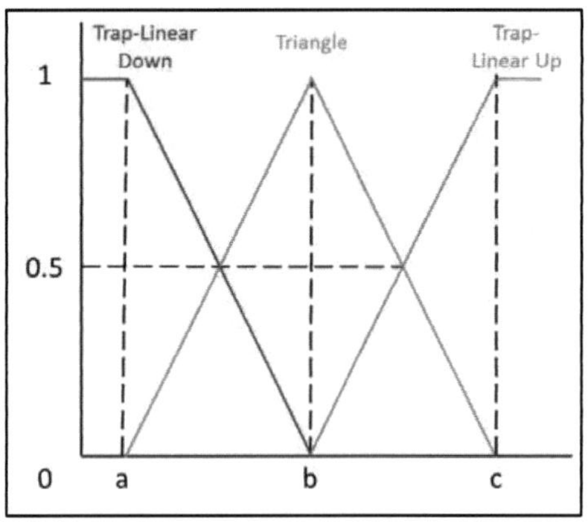

Figure 2. The Membership Curves of Fuzzy Output

3.2. Data Inference

Fuzzy inference strongly relies on the fuzzy rules and the membership degree of the fuzzy sets, i.e. to apply the implication function and compose all the fuzzy outputs. In this stage, we applied two levels of data reasoning. First, we applied the monotonous reasoning method for each of main environment indicators, i.e. to define the index of river quality, air quality and the coverage of land cover. Second, we combined these indexes to represent the index of Environment Quality Performance (EQP) by using the min-max inference approach.

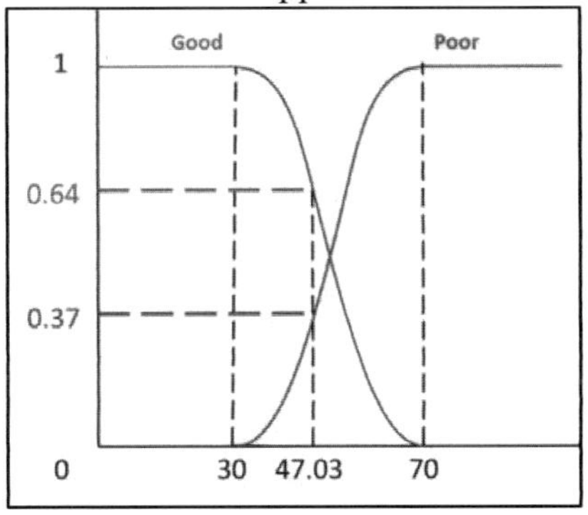

Figure 3. Fuzzification of TSS

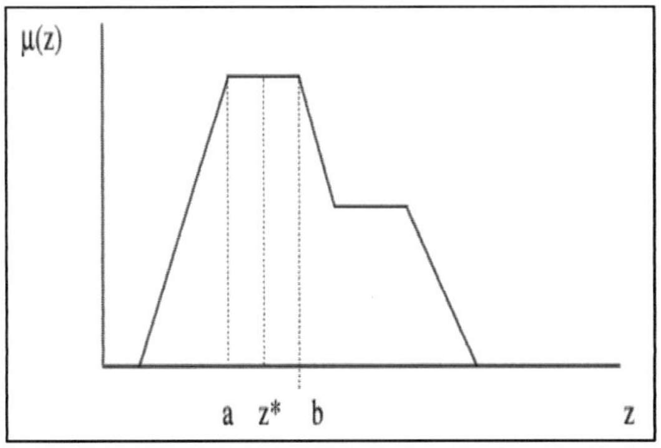

Figure 4. Mean Of Maximum (MOM) Defuzzification

On the first method, the monotonous reasoning, we related two fuzzy variables with simple implication rules. E.g. rule for determining a "High" level of river quality index was defined as follows:

IF river quality is Good THEN river index is High

The value of river quality was obtained based on the average of all the components forming the index, i.e. the average of "Good" membership degree from seven parameters (TSS, DO, COD, BOD, TP, FC, TC) and mapped it to the "High" membership function to represent the river index. Similar approaches were also applied to define the "Low" river index based on the average of "Poor" values of all river parameters, and to define the "High" and the "Low" index of the air quality and the coverage of land cover.

Table 1. Formation of Fuzzy Rules

Rules #	**Antecedents**			**Consequent**
	River	*Air*	*Land Cover*	*EQP*
1	High	High	High	High
2	High	High	Low	Moderate
3	High	Moderate	High	High
4	High	Moderate	Low	Moderate
5	Low	High	High	Moderate
6	Low	High	Low	Moderate
7	Low	Moderate	High	Moderate
8	Low	Moderate	Low	Low

On the second reasoning method, we applied the minimum function to determine the implication of fuzzy rules and aggregated all output based on the maximum function. In this step, we had defined 8 fuzzy rules which have been derived from all combinations of environment indicator fuzzy sets as the rule's antecedent, in which each indicator consisted of two fuzzy sets to determine the general index of Environmental Quality Performance (EQP) as the rule's consequent as seen in Table 1. The first rules can be stated as follow:

IF (river index is High and air index is High and land cover is High) THEN EQP is High

3.3. Defuzzification

Defuzzyfication is a process to obtain the crisp value as the representation of fuzzy natures. There are various methods to calculate the crisp output of fuzzy system. The most common method is the centroid function (composite moment). However, in this study we applied the Mean of Maximum (MOM) which is simple in computation and is able to give the reliable results. In MOM, the crisp value (z*) is based on the average values of fuzzy sets with the maximum of membership degree as shown in Fig. (4) and Eq. (6).

$$z^* = \frac{a+b}{2} \tag{6}$$

where:
 a = bottom boundary
 b = upper boundary

4. Results and Discussion

The model for indexing the quality of environmental performance based on fuzzy inference approach in 33 provinces in Indonesia is presented here. We applied the environmental condition data set that was published by the Ministry of Environment of Indonesia in 2015.

In the beginning, we analyzed the result of fuzzification step. We compared the membership degree of 10 environmental quality parameters of 33 provinces. We used the "Good" membership degree which had the value more than 0.5 since it may represent data within the safe limits. As seen in Fig. (5), we inspected that of these 10 parameters the best circumstance was found in SO2 in which all provinces in Indonesia were within the limits of healthy air.

Meanwhile the lowest circumstance was found in the Land Cover (LC) since the number of provinces with the coverage of forested area larger than 40%, was smaller than its opposite side and had the smallest value of all parameters. Especially for the river index parameters, i.e. DO, TP, COD, TSS, FC, TC and BOD, we found the lowest circumstance in Biochemical Oxygen Demand (BOD).

We also analyzed the spatial distribution of these "Good" membership degrees. To simplify the explanation, we defined 5 groups of data based on the most well-known islands, i.e. Java, Kalimantan (abbreviated as Kali), Nusa Tenggara Papua (abbr. as Nuspa), Sulawesi (abbr. as Sula) and Sumatra (abbr. as Suma). The list of group members is shown in Table 2.

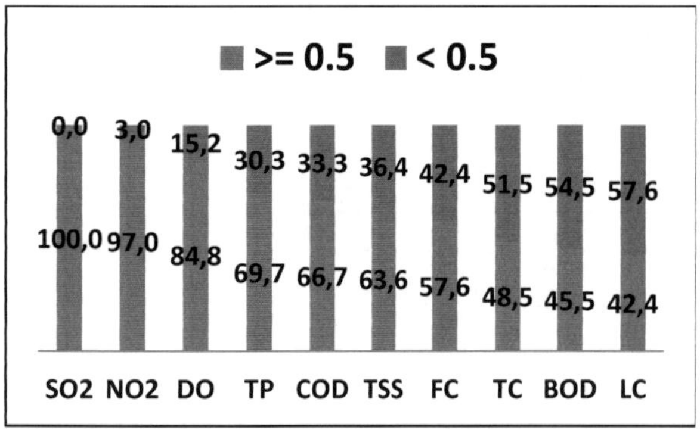

Figure 5. Distribution of GOOD Membership Degree

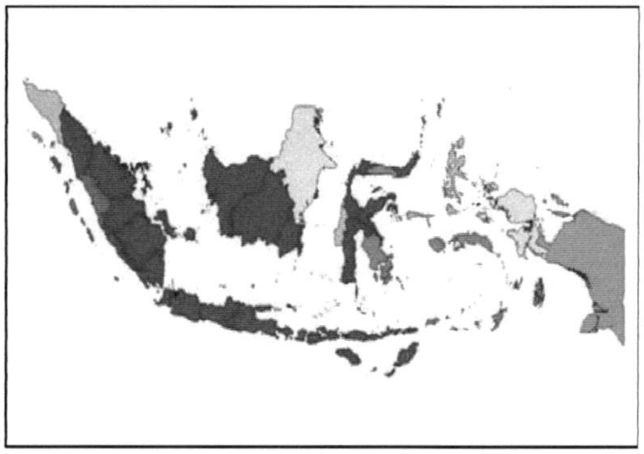

Figure 6. Map of EQP Distribution in Indonesia

In more detail, we analyzed the spatial distribution of island entities with "Good" membership degree below 0.5. This analysis was done for determining what indicators should have more consideration by these five major islands. As seen in Table 3, Java had TSS value of 50% indicating that 3 of 6 provinces had poor circumstances since their TSS were above the allowed threshold. Similarly, we inspected that

Java also had the poor circumstances on COD, BOD, TC, FC and LC. Kalimantan had the poor circumstances on BOD, Nusa Tenggara Papua (Nuspa) had the poor circumstances on BOD and TC, Sulawesi had the poor circumstances on TSS, and Sumatra had the poor circumstances on the coverage of land cover (LC).

Afterwards, we analyzed the result of data inference based on the monotonous and min-max reasoning. We found that the highest index of "High" river quality performance was 95.88 (in the scale of 0 to 100), including a province in Sumatra, and the lowest index was 10.73 which was located in a province in Java. For the index of air quality performance, we found that most provinces were having the value above 75, except one province in Java which had the air quality index below 50. For the land cover, we found the highest index was 100 which were located in 4 provinces in Nuspa and one province, respectively, in Kalimantan and Sulawesi. Meanwhile for the lowest index of land cover it was 0 and found in 8 provinces which dominantly were located in Java and Sumatra Island.

At the end, we analyzed the defuzzification results. The main output of defuzzification process was the environmental quality index of each province in Indonesia in the range of value from 0 to 100. In this final step, we found that 10 provinces had the EQP above 80 with the dominance of provinces in the group of Nusa Tenggara Papua (Nuspa) and none were located in Java Island. Meanwhile the lowest EQP was at 11.04 in a province in Java Island. The rest of 22 provinces had the EQP values between 50 and 80. To visualize the spatial distribution of EQP value among these 33 provinces in Indonesia, we illustrated them as a map as seen in Fig. (6), i.e. a darker red color indicated a lower value of EQP.

Table 2. List of Island Members

Island	Member of Province
Java (6)	Jakarta, West Java, Central Java, Yogyakarta, East Java, Banten
Kali (4)	West Kalimantan, Central Kalimantan, South Kalimantan, East Kalimantan
Nuspa (7)	Bali, West Nusa Tenggara, East Nusa Tenggara, Maluku, North Maluka, Papua, West Papua
Sula (6)	North Sulawesi, Central Sulawesi, South Sulawesi, South East Sulawesi, West Sulawesi, Gorontalo
Suma (10)	Aceh, North Sumatra, West Sumatra, Riau, Island of Riau, Jambi, South Sumatra, Bengkulu, Lampung, Bangka Belitung

Table 3. Percentage of Provinces with Good Membership Degree < 0.5

Params	Java	Kali	Nuspa	Sula	Suma
TSS	50.0	25.0	14.3	83.3	20.0
COD	66.7	25.0	14.3	33.4	30.0
DO	33.4	0	0	33.4	10.0
BOD	100	75.0	57.2	16.7	40.0
TP	50.0	0	42.9	16.7	30.0
TC	100	50.0	57.2	50.0	20.0
FC	100	50.0	28.6	33.4	20.0
SO2	0	0	0	0	0
NO2	16.7	0	0	0	0
LC	100	50.0	14.3	33.4	80.0

5. Conclusions

Based on the research findings, there are several conclusions that can be drawn here:

1. On the fuzzification step, there were analyses regarding environmental parameter circumstances. We found that among three types of index that form the Environmental Quality Index, the best circumstances was the air quality performance, meanwhile the coverage of the green land cover was the lowest circumstance in Indonesia.

2. Data inference has been done based on the monotonous and mix-max reasoning to quantify the performance of each indicator which was part of the environmental quality index. Meanwhile, the control rules as the center of data inference were derived from all combinations of environment indicator fuzzy sets. However, these rules only cover the three main indexes, i.e. river, air, and land cover not included the environmental parameters. Hence, we still need a better method to generate the fuzzy rules which also consider all aspects of environmental performance.
3. We also have provided an analysis of data spatial distribution, i.e. research objects were grouped into five islands, and found that Java Island had the most number of provinces with the environmental parameters below the threshold or the safe limits.
4. This research showed that the environmental quality performance can be quantified without eliminating the natures of the data set and is simultaneously able to show the environment behavior by considering the control rules as the representation of human way of thinking and reasoning along with its spatial pattern distribution.

Acknowledgment
Thank you to the Directorate of Research and Service Communities of Universitas Islam Indonesia for the funding and also for Hadi Kusuma Hidayat for providing the data sets.

References
[1] Streimikiene D 2015 Environmental indicators for the assessment of quality of life *Intellect. Econ.* **9** pp 67–79.
[2] Mei N S, Wai C W, and Ahamad R 2016 Environmental awareness and behaviour index for Malaysia *Procedia - Soc. Behav. Sci.* **222(7)** pp 668–75.
[3] Silva L T 2015 Environmental quality health index for cities *Habitat Int.* **45**(P1) pp 29–35.
[4] Environment M 2012 *Report of Environmental Quality Index in Indonesia 2012 (Indeks Kualitas Lingkungan Hidup Indonesia).*
[5] Environment M and Forestry M 2014 *The Statistics of Enviromental Quality and Forestry in Indonesia 2014 (Statistik Kementrian Lingkungan Hidup dan Kehutanan Tahun 2014).*
[6] Kang Y K, Kim H, Heo G, and Song S Y 2017 Diagnosis of feedwater heater performance degradation using fuzzy inference system *Expert Syst. Appl.* **69** pp 239–46.
[7] Effendi H 2016 River water quality preliminary rapid assessment using pollution index *Procedia Environ. Sci.* **33** pp 562–7.
[8] Alavi N 2012 Using Mamdani fuzzy inference system to quality determination of Mozafati dates *J. Saudi Soc. Agric. Sci.* **12(2)** pp 137–42.
[9] Guillaume S and Charnomordic B 2012 Fuzzy inference systems: An integrated modeling environment for collaboration between expert knowledge and data using FisPro *Expert Syst. Appl.* **39(10)** pp 8744–55.

Study of Colour Model for Segmenting Mycobacterium Tuberculosis in Sputum Images

A Kurniawardhani, R Kurniawan, I Muhimmah, and S Kusumadewi

Informatics Department, Islamic University of Indonesia, Yogyakarta, Indonesia.

Abstract. One of method to diagnose Tuberculosis (TB) disease is sputum test. The presence and number of Mycobacterium tuberculosis (MTB) in sputum are identified. The presence of MTB can be seen under light microscope. Before investigating through stained light microscope, the sputum samples are stained using Ziehl–Neelsen (ZN) stain technique. Because there is no standard procedure in staining, the appearance of sputum samples may vary either in background colour or contrast level. It increases the difficulty in segmentation stage of automatic MTB identification. Thus, this study investigated the colour models to look for colour channels of colour model that can segment MTB well in different stained conditions. The colour models will be investigated are each channel in RGB, HSV, CIELAB, YCbCr, and C-Y colour model and the clustering algorithm used is k-Means. The sputum image dataset used in this study is obtained from community health clinic in a district in Indonesia. The size of each image was set to 1600x1200 pixels which is having variation in number of MTB, background colour, and contrast level. The experiment result indicates that in all image conditions, blue, hue, Cr, and Ry colour channel can be used to segment MTB in one cluster well.

1. Introduction

Tuberculosis (TB) is included the top 10 disease that cause death. In 2015, there were 1.8 million people in the world died from TB, in other words, there are over 4,900 people died every day [1]. Indonesia is one of the top three high burden country for TB in the world. In 2015, there were 1.02 million TB incidence in Indonesia [2]. These large number of TB suffers can be caused of TB, one of communicable disease, is spread easily through splashes of sputum in the air that contain Mycobacterium tuberculosis (MTB), while the sufferer cough, spit, or sneeze. Moreover, to identify that a person has TB takes a long time. For decreasing the number of TB suffers, especially in low- and middle-income countries, affordable, reliable, and fast enough instrument is necessary.

More than 95% of deaths caused by TB occur in low- and middle-income countries [1]. Diagnostic test commonly used in low- and middle-income countries is examination of sputum samples [1]. Sputum samples were examined through a microscope to see the presence of MTB, name of the bacteria that causes TB. Because inexpensive and easy to maintain, the light field microscope is used commonly in low- and middle-income countries. Before sputum samples are seen under light microscope, the sputum sample is stained using Ziehl–Neelsen (ZN) stain technique.

To develop affordable, reliable, and fast enough instrument, a lot of studies are conducted. There are many studies that start by developing automatic MTB detection in sputum image using light field microscope and ZN staining [3-12]. Those studies were attempting for assisting pathologists in assessing sputum sample to diagnosis TB. The studies have major challenges, namely, sputum images have inhomogeneity intensity of background, number of MTB, and contrast between one image with another images, as shown in Figure 1. When the ZN staining process is done properly, MTB will appear in red colour while other organisms and background will in blue colour [12], and the samples will have adequate contrast, as shown in Figure 1.a. Improper staining procedure and reagent preparation can produce under-stained or over-stained images [13-14], as shown in Figure 1.b and Figure 1.c, respectively. Under-stained images make images have low contrast, while over-stained images make images have red background and dark red, red, or bright red MTB.

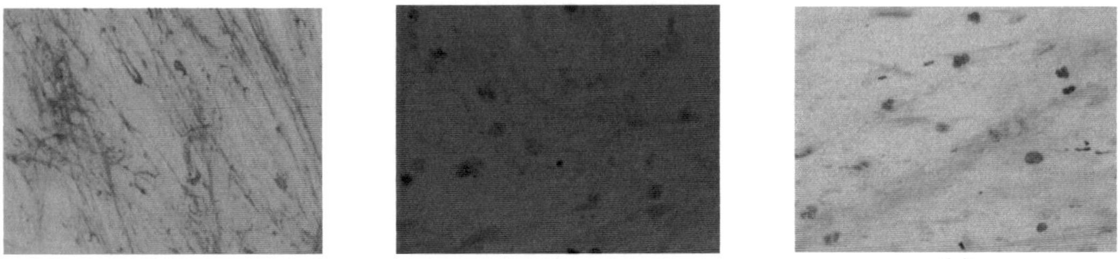

(a) (b) (c)

Figure 1. The sputum images. (a) Normal images. (b) Under-stained image. (c) Over-stained image.

The staining issue as previously mentioned increase the difficulty in MTB segmentation process. To overcome that problem, different colour models are considered used in segmentation process. In [3-8], segmentation process is done in RGB (Red Green Blue) colour model. RGB colour model that is adopted from physiology of the human eye is renowned colour model which is ubiquitous used in digital instrument. It, however, cannot mimic the human colour visual system appropriately. In those studies, MTB can be segment well because most of sputum images used in those studies have pretty fine stained. In [9-10], segmentation process is done in HSV (Hue Saturation Value) colour model, which segmenting MTB based on the degree value in Hue channel. HSV that is adopted from how choose colour for paint or ink can defined colour in similar way with how humans perceive colour. Nevertheless, it overlooks the detail of colour appearance. In [11], segmentation process is done in CIELAB (Commission Internationale de l'Eclairage Lab) colour model. CIELAB which is 'device-independent' has gamut wider than human vision. In [12], segmentation process is done not only in CIELAB (Commission Internationale de l'Eclairage Lab) colour model, but also combined it in YCbCr colour space. Meanwhile, [13,14] do segmentation process in RGB and C-Y colour model. Because of many variation colour model used in segmentation proses, this study will investigate the appropriate colour model used in segmentation process that can apply in different stained conditions. The colour models will be investigated are each channel in RGB, HSV, CIELAB, YCbCr, and C-Y colour model.

2. Image Dataset

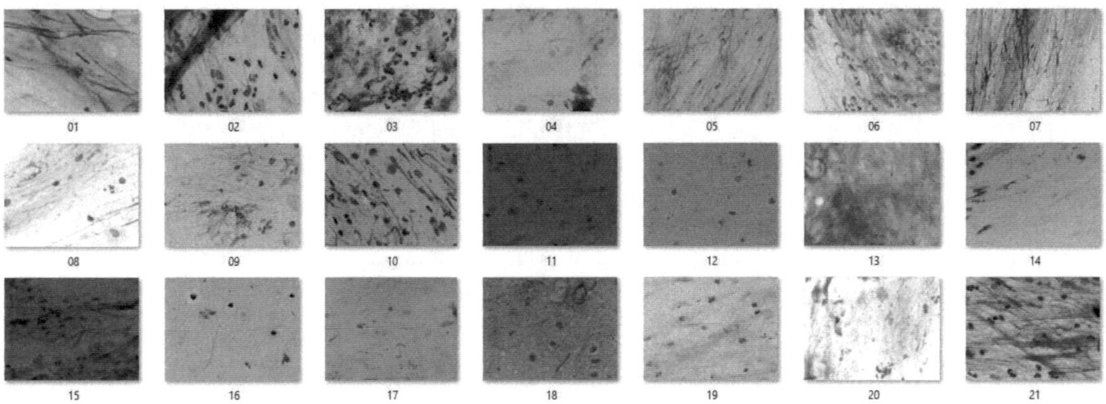

Figure 2. The examples of sputum image database.

Sputum microscopic images of Tuberculosis (TB) sufferer are collected from three community health clinic (CHC) (called Puskemas in Indonesia) in Temanggung Regency, Central Java, Indonesia. The three CHC are Bansari, Ngadirejo, and Traji Puskesmas. Temanggung Regency was chosen because that area has a fairly high TB patient so that has many digital collections of sputum microscopic images of TB patients. From those three CHC, the 234 sputum images of TB patients were obtained. All images contained Mycobacterium tuberculosis (MTB) with varying amounts in each image. Each image has different colouring and lighting conditions against to another images. The sputum image is *. TIF format and 24-bit colour image. The size of each image was set to 1600x1200 pixels. The examples of sputum images used are shown in Figure 2. Those sputum image database with specification of Indonesian people can be accessed by public for research and education purpose at http://psimed-zinsmid.informatics.uii.ac.id/.

3. Colour Model

Colour model is numerical value representation of colour to model visible spectrum of colour. Meanwhile colour space is a variant of colour space that specifies a gamut (range) of colours [16].

3.1. RGB (Red Green Blue)

The Red Green Blue (RGB) colour model is commonly colour model used in digital instruments, such as scanner, camera, digital TV, computer, and mobile phone. Red Green Blue means that it colour model have three primary colours, namely, red, green, and blue. To create another colour besides of the three basic colours is done by combining three basic colours in accordance with the desired colour composition. The principle used in RGB is adopted from physiology of the human eye [15].

3.2. HSV (Hue Saturation Value)

The Hue Saturation Value (HSV) colour model is adapted from how people choose colour for paint or ink. Because of that, it claimed that HSV can defined colour in similar way with how humans perceive colour. Hue describes the colour or tint information, saturation describes how white the colour is, and value describes how dark the colour is [15]. To get HSV colour, RGB image can be convert to HSV using equation (1-3).

$$H = \cos^{-1}\left(\frac{\frac{1}{2}\big((R-G)+(R-B)\big)}{(R-G)^2+(R-B)+(G-B)}\right)^{\frac{1}{2}} \tag{1}$$

$$S = 1 - \frac{3}{(R+G+B)}\min(R,G,B) \tag{2}$$

$$V = \frac{R+G+B}{3} \tag{3}$$

3.3. CIELAB (Commission Internationale de l'Eclairage Lab)

CIELAB consist of three-dimension colour representation, namely, ***L*** for luminance or lightness information, ***a*** for colour opponent information between red and green, and ***b*** for colour opponent information between blue and yellow.

3.4. YCbCr

The YCbCr colour space is commonly used for digital video. ***Y*** describes the luminance or lightness information, while ***Cb*** and ***Cr*** describes chrominance information. ***Cb*** represent the colour difference between blue-component and a reference value, while and ***Cr*** represent the colour difference between red component and a reference value. To get HSV colour, YCbCr image can be convert to YCbCr using equation 4).

$$\begin{bmatrix} Y \\ R_y \\ B_y \end{bmatrix} = \begin{bmatrix} 0.299 & 0.587 & 0.114 \\ -0.1687 & -0.3313 & 0.5 \\ 0.5 & -0.4187 & -0.0813 \end{bmatrix} \begin{bmatrix} Red \\ Green \\ Blue \end{bmatrix} \quad (4)$$

3.5. C-Y colour

The C-Y consist of three-dimension colour representation ***Y, Ry,*** and ***By***. ***Y*** describes the luminance or lightness information, while ***Ry*** and ***By*** describes chrominance information. ***Cb*** represent the colour difference between blue-component and a reference value, while and ***Cr*** represent the colour difference between red component and a reference value. To get C-Y consist, RGB image can be convert to YCbCr using equation (5).

$$\begin{bmatrix} Y \\ R_y \\ B_y \end{bmatrix} = \begin{bmatrix} 0.299 & 0.587 & 0.114 \\ 0.701 & -0.587 & -0.114 \\ -0.299 & -0.587 & 0.886 \end{bmatrix} \begin{bmatrix} Red \\ Green \\ Blue \end{bmatrix} \quad (5)$$

4. Method

4.1. K-means Clustering

K-means clustering is one of well-known clustering method that has been widely used in various fields. K-means clustering is famous for its simplicity and reliability. K-means clustering divides the data into multiple clusters, with the number of clusters corresponding to the k value specified by the user [17]. The algorithm of K-means clustering is:

1. Determine the value of k
2. Determine the cluster centre coordinates as much as k, randomly
3. Calculate the distance of each data point with each cluster centre
4. Determine the membership of each data point against a cluster based on the nearest distance between the data points to one of the cluster centres.
5. Recalculate the cluster centre based on the average value of all data point of cluster members
6. Do step 3-5 continuously until the coordinates of cluster centre does not change.

4.2. Segmentation Process

The segmentation process aims to separate objects and backgrounds. Objects in the sputum image not only Mycobacterium tuberculosis (MTB), but also can be cell remnants, food scraps, fibres, or blood. This segmentation process will look for the colour model channel which can segment most of MTB against to the background and all non-MTB objects. The scheme of the segmentation process can be seen in Figure 3.

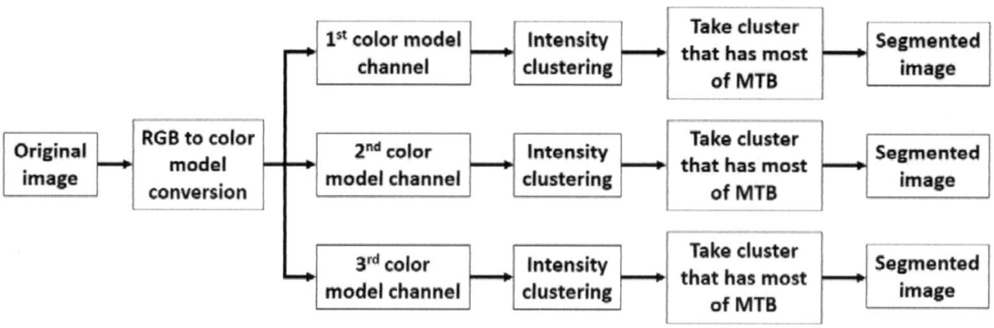

Figure 3. The scheme of the segmentation process.

Figure 3. shows the stages of the segmentation process of each colour model which each colour model has three colour channels. Because there are five colour models that will investigated in this paper, namely, RGB, HSV, CIELAB, YCbCr, and C-Y colour model, the scheme that shows in Figure 3 will be repeated five times. Therefore, the 15 segmented images will be gotten after this step, namely, segmented image in Red, Green, Blue, Hue, Saturation, Value, ***L***, ***a***, ***b***, ***Y***, ***Cb***, ***Cr***, ***Y***, ***Ry***, and ***By*** colour channel.

The first stage of segmentation proses is conversion. The original images gotten from local Community Health Clinic (CHC) are in 24-bit colour of RGB. To investigate those images in HSV, CIELAB, YCbCr, and C-Y colour model, the colour conversion is necessary. After the five images in the five colour models are obtained, each colour model images are separated based on their three colour model channels. Therefore, there are 15 colour channel images is derived from one sputum image.

Clustering process using k-Means algorithm is performance at each colour channel of 15 colour channel images. K-Means Clustering is used to cluster the pixels to be one cluster if the pixels have same colour and separate them to different cluster if the pixels have the different colour. Hence, a cluster will have object that have similar colour. The parameter of ***k*** for k-Means clustering used in this study is five, so that there are five clusters produced. Further, the cluster that contains most of MTB objects is selected for the result of segmentation process.

5. Result and Discussion

The experiments are conducted in accordance with stages that shown in Figure 3 and used sputum images that have variation in number of MTB, background colour, and illumination. Firstly, the best ***k***-parameter of k-Means clustering used in this study is investigated. Secondly, how to select the cluster that obtained from k-Means clustering is considered to get the best segmented images from each colour channel. Thirdly, the segmented images are investigated to seek the best representative colour channel in MTB segmenting process. Hardware specifications used in this study are Intel® Core™ i5 Processors and 4 GB RAM. We run the experiment in Microsoft Windows 10 Professional x64. All experiment is performed using Matlab 2013.

For selecting k-parameter for k-Means clustering, the 15 images that have variation in number of MTB, background colour, and illumination are selected randomly. The value of ***k***-parameter is varied from two up to seven. The experiment results define that the average of optimum value for ***k***-parameter for 225 images, 15 images multiplied by 15 colour channel, is five. It means in an images the colour will grouped into five clusters. It is due to in an image there are various objects, such as fibber, cell remnants, food scraps, and blood. Although they have the same colour properties, but they have different colour gradation level because of improper staining process. MTB sometimes have the same colour with other objects around them or even similar to the one of gradation colour level of background. Thus, the level of gradation also needs to be grouped again.

For selecting the cluster that obtained from k-Means clustering, the 15 images that have variation in number of MTB, background colour, and illumination are selected randomly too. In order to get the best segmented images from each colour channel, an image in each cluster is investigated and compared to other images that belongs to remaining clusters. In most of experiment, most of MTB can be segmented in a cluster that have the fewest pixel number of members. When the ZN staining process is done properly, MTB will appear in red colour while other organisms and background will in blue colour. Moreover, MTB has tiny size about 0.3-0.6 μm in width and 1–4μm in length, which give a size of 70–140 pixels in the image [10,14]. In some image, however, MTB cannot be clustered alone because the red colour of MTB does not look conspicuous. It is due to the image has low contrast or MTB are blocked by another object. Furthermore, when the ZN staining process is done improperly, background can appear in bright red colour, which is almost the same colour as MTB.

The colour channel lists that can segment MTB well are shown in Table 1. In all image condition, blue, hue, Cr, and Ry colour channel can be used to segment MTB in one cluster well. It is due to blue colour channel have different intensity in MTB and background colour. In background colour, which background colour tends to white either in the red or blue background, blue intensity tends to be higher than it is in MTB colour. On the other side, red intensity tends to have similar intensity for MTB and background colour. MTB colour tends to in red, magenta, or purple colour, that have combination colour between red and blue. In some of images, however, the segmented image in blue colour channel contain another blue object that should be eliminated. Hue describes the colour or tint information. Thus, Hue can segment the colour purity information which is not affected by brightness information. In most of images, however, the segmented image in hue colour channel contain a tiny object like noise that should be eliminated. Cr and Ry colour channel have the similar segmented result. In Cr and Ry colour channel, most of MTB can be grouped as one cluster but MTB is not separated from background. MTB colour properties in Cr and Ry colour channel tends to have similar colour properties of bright background, either in the red or blue background. The examples of segmented images are shown in Figure 4.

Table 1. List of colour channel that can segment MTB.

Image	Background Colour	Contrast Level	Number of another object	Number of MTB	Colour Channel
01.TIF	Blue	Low	±20	8	Blue, hue, value, Cr, Ry, a, green
03.TIF	Blue	Low	>70	±50	Hue, Cr, Ry, blue, value
14.TIF	Bright blue	Low	±20	10	Blue, green, hue, a, value, luminance, Y, Cr, Ry
08.TIF	White	High	±20	13	Hue, a, value, green, blue, Cr, Ry
16.TIF	Magenta	High	±10	±15	a, Cr, Ry, luminance, green, saturation, blue, value, hue
07.TIF	Bright red	High	>50	4	Blue, a, saturation, hue, Cr, Ry, green
17.TIF	Bright red	Low	±20	18	a, Cr, Ry, saturation, blue, green, hue
18.TIF	Bright red	Low	±20	>70	a, hue, Cr, Ry, blue, green

The 03.TIF is little bit difficult to segment because it has many pretty big dark blue objects and tiny MTB. Many MTB in that picture are blocked by pretty big dark blue objects. Hence, further study is necessary in order to address some issues, such as image enhancement before doing segmentation because most of the images have low contrast.

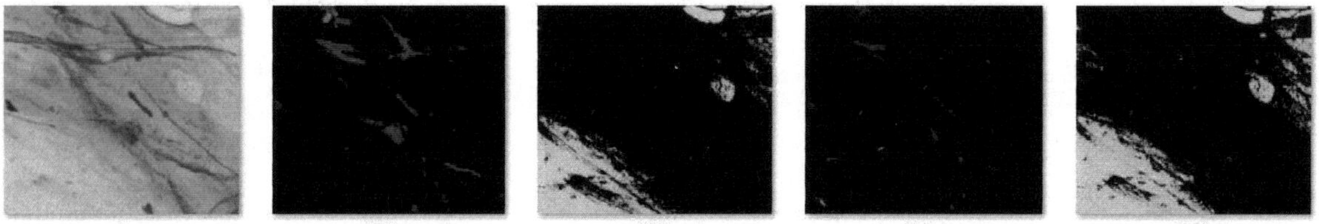

Figure 4. The examples of segmented images.

6. Conclusion

This study investigated the colour models to look for colour channels of colour model that can segment MTB well in different stained conditions. The colour channels investigated are each channel in RGB, HSV, CIELAB, YCbCr, and C-Y colour model. K-Means clustering algorithm is used to segment the colour information. The sputum image dataset used in this study has variation in number of MTB, background colour, and contrast level. The experiment result indicates that in all image conditions, blue, hue, Cr, and Ry colour channel can be used to segment MTB in one cluster well.

References

[1] World Health Organization (WHO) 2017 *Fact Sheets–Tuberculosis* [Online] Available: http://www.who.int/mediacentre/factsheets/fs104/en/
[2] World Health Organization (WHO) 2017 *Global tuberculosis report 2016* [Online] Available: http://www.who.int/tb/publications/global_report/en/
[3] Costa M G F, et al 2008 Automatic identification of mycobacterium tuberculosis with conventional light microscopy *Eng. in Medicine and Biology Society (EMBS)* pp 382-5.
[4] Sadaphal P, Rao J, Comstock G W, and Beg M F 2008 Image processing techniques for identifying Mycobacterium tuberculosis in Ziehl-Neelsen stains *The Int. J. of Tuberculosis*

and Lung Disease, 12(5), p 579-582.
[5] Khutlang R, Krishnan S, Whitelaw A, and Douglas T S 2009 Detection of tuberculosis in sputum smear images using two one-class classifiers *Biomedical Imaging: From Nano to Macro 2009 (ISBI'09) IEEE International Symposium on* pp 1007-10.
[6] Khutlang R, Krishnan S, Dendere R, Whitelaw A, Veropoulos K, Learmonth G, and Douglas T S 2010 Classification of Mycobacterium tuberculosis in images of ZN-stained sputum smears *IEEE Transactions on Information Technol. in Biomedicine* **14(4)** pp 949-57.
[7] Mukti J and Kale K V 2011 Analysis of zn-stained sputum smear enhanced images for identification of mycobacterium tuberculosis bacilli cells *Analysis* **23(5)**.
[8] Raof R A A, Mashor M Y, Ahmad R B, and Noor S S M 2011 Image segmentation of Ziehl-Neelsen sputum slide images for tubercle bacilli detection *Image Segmentation InTech.*
[9] Xu C, Zhou D, Guan T, and Liu Y 2014 A segmentation algorithm for mycobacterium tuberculosis images based on automatic-marker watershed transform *Robotics and Biomimetics (ROBIO) IEEE International Conference on* pp 94-8.
[10] Makkapati V, Agrawal R, and Acharya R 2009 Segmentation and classification of tuberculosis bacilli from ZN-stained sputum smear images *Automation Sci. and Eng. IEEE Int. Conf. on* pp 217-20
[11] Rulaningtyas R, Suksmono A B, Mengko T, and Saptawati P 2015 Multi patch approach in K-means clustering method for color image segmentation in pulmonary tuberculosis identification *Instrumentation, Communications, Information Technol., and Biomedical Eng. (ICICI-BME)* pp 75-8.
[12] Sotaquira M, Rueda L, and Narvaez R 2009 Detection and quantification of bacilli and clusters present in sputum smear samples: a novel algorithm for pulmonary tuberculosis diagnosis *Digital Image Processing* pp 117-21.
[13] Osman M K, Mashor M Y, and Jaafar H 2010 Detection of mycobacterium tuberculosis in Ziehl-Neelsen stained tissue images using Zernike moments and hybrid multilayered perceptron network *Systems Man and Cybernetics (SMC)* pp 4049-55.
[14] Osman M K, Mashor M Y, Saad Z, and Jaafar H 2010 Colour image segmentation of tuberculosis bacilli in Ziehl-Neelsen-stained tissue images using moving k-mean clustering procedure *Mathematical/Analytical Modelling and Computer Simulation (AMS)* pp 215-20.
[15] Gonzalez R C and Woods R E 2007 *Digital Image Processing* 3 ed (Prentice Hall).
[16] Lal S 2007 *Photoshop CS* (New Delhi: Laxmi Publication Pvt. Ltd).
[17] MacQueen J June 1967 Some methods for classification and analysis of multivariate observations *Proc. of the fifth Berkeley Symp. on Math. Statistics and Probability* **1(14)** p 281-97.

Acknowledgments
Authors wishing to acknowledge to DRPM Director General of Ministry of Research, Technology and Higher Education (Kemenristek Dikti) who has provided financial support through Prime Development of Higher Education research scheme in 2017 so that the research of "Prototype of TB Detection System based on Sputum Image" can be conducted.

IOP Conf. Series: Materials Science and Engineering **325** (2018) 012011 doi:10.1088/1757-899X/325/1/012011

The Role of Participatory Design in Mobile Application Development

Almed Hamzah

Department of Informatics, Universitas Islam Indonesia
Jalan Kaliurang KM 14,5 Sleman Yogyakarta Indonesia 55584

almed.hamzah@uii.ac.id

Abstract. Mobile devices are used by people worldwide. It becomes a common equipment to complete a day-to-day activity. Inside the devices, there are numerous mobile applications that have been built for various needs. Some of these are quite successful while the other are not. The development of successful mobile application faces several challenges. In this research, we want to explore the use of participatory design method in mobile application development. Particularly, the aim of the study is to answer the question whether participatory design method has a place in the realm of mobile application development. We established two sessions of workshop to accommodate the participant to take part in the development process of mobile application. The result shows that participatory design method can determine how the user will deal with the limitations of mobile devices. It helps user to create a particular form of interaction that meets mobile devices characteristics.

Ke ywords: participatory design, mobile application, exploratory study

1. Introduction

Mobile applications are flourishing along with the increase of mobile devices user itself. Recently, there are about 2 billion mobile devices being deployed [1]. With the device, people are able to download and use various mobile applications which currently reach more than two million different applications [2]. This number is far more than the number of application available five years ago which only around 200 thousand applications [3].

However, the development of mobile applications face several challenges such as information architecture, data management and application context, and user interface [4]. This challenges are mostly stemmed from technical limitations of mobile devices [5]. To some extent, the challenges will affect the usability of mobile applications. In some cases, the user even unable to solve their problem with the applications that supposed to help them in that problem because of lack usability.

Participatory design is somewhat a new approach to the development of mobile application paradigm. This method is actually the mechanism of involving the user since the beginning until the end of the development of the mobile application. The user has a freedom to explore and do experiment with various applications' features. Nonetheless, the user still need accompaniment from the designer of application so that the development process is appropriately running.

This research aims to explore the role of participatory design method in the development of mobile applications. However, the biggest challenges in using this method is how to encourage user, as a partner of design, to be able to actively express their idea to the designer of application [6], [7] and then manifest the idea into sketch, prototype, or storyboard. In addition, the lesson learnt is important for the other designer whom build the same application in similar context [8], [9].

The next section will describe another study with similar topic. It then followed with the explanation of methodology that have been used in this study. After that, the result will be presented followed by discussion section. The conclusion section will end this paper.

2. Related Works

It seems that participatory design has been considered as useful approach in producing highly usable application that match users' needs in a better way. For instance, there is a report about a work that combining participatory design approach in agile development [10]. The use of method makes for better balance of user's involvement and a tight schedule that innate in the development process.

However, participatory design is highly dependent on the context where it is implemented [8]. To some extent, it is not interchangeable. This is due to the heterogeneity of user themselves which mostly differ from one to another background. For instance, people coming from medical background is somehow different with those who coming from academic background in terms of how they do thinking and or doing something which in turn will affecting the design process and the application itself. Some researchers have been studied the use of participatory design in healthcare context. It helps them to find unique requirements and limitation of design [9] that was strongly correlated to the condition in which the system was implemented [8]. Moreover, it can be used to take advantage of user existing knowledge [11].

In the same manner, mobile application development is also a context-dependent activity as stated by Harpe [8] that it is affected by the context where the application would be deployed. Therefore, the research that related to the development of mobile application in various context is important in order to take part in the future research or development that met the similar context.

3. Method

This study is explorative and cross-sectional in nature. The research route begins with literature review to find out the way to explore participatory design approach in mobile application development. Then, it was followed by two workshop sessions and ended up with analysis. There are n=100 participants joining this research. They work in a group consisting of 4 to 5 people. They work in the following steps.

Firstly, they were interviewed about the application they want to build. As a case study, we choose to build mobile-based learning management system as a big theme in this research. It was perfectly match with the day-to-day activity of participants since all of them are students.

After that, we held two workshop sessions to accommodate the development process with the participant.

1. Workshop 1: storyboarding. The participants would explore their idea in a group and create some storyboard.
2. Workshop 2: prototyping. The participants would develop high-fidelity prototype based on the idea that have been stored in storyboards.

Each workshop sessions were followed by discussion with another group to retrieve feedback.

After the prototype was formed and fixed, we hire professional programmer to build the applications based on participants' prototype. We built the first version of application for Android platform. This is merely due to the popularity of this platform among participants.

4. Results

Most of literature have been reviewed stated that their research is important to other researcher or designer in terms of giving an example of either the implementation of participatory design or the development of mobile solution. It can be concluded that there are still a lot of work to be done in this area so that there is enough example in various context. Therefore, this research proposes the context in academic nature.

In the first workshop session, participants were brainstorming their idea about the features and appearance of their forthcoming applications. In this workshop, they work as a group consist of 5 participants. Along with that, we give them a slight review of some well-known similar application, such as Moodle, Classroom (Google), and Edmodo. In addition, we explain to the participant how to build a low-fidelity prototype by sketching and storyboarding. The goal is to collect user requirements as much as possible. Participants are free to explore their idea and put it in the storyboard.

After the participant had finished their storyboard then we held presentation and feedback session. In this session, each group were presenting their storyboard to the other groups and taking feedback from them and this is the end of the first workshop session. Through the workshop, we had identified 20 most preferable features of mobile application that participants want to build. Table 1 below depicts the features together with the number of groups who brings the idea and the percentage of that groups comparing to the whole groups of participants.

From the list in Table 1, we do not build all the features. Instead, we choose to highlight (bold printed) and finally built only several features due to the limitation of resources (personnel and time). The selection is based on the observation during the workshop. The selected features are the features that discussed intensely by the participants and deeply attracted them.

Table 1 Preferable features

	Features	Number of Groups	Percentage
1	Assignments	17	89,47
2	Discussion/Forum	11	57,89
3	Course Material Sharing	11	57, 89
4	Login / Authentication	10	52,63
5	Schedule	9	47,37
6	Grade	7	36,84
7	Notification	6	31,58
8	Message	6	31,58
9	Pair View	5	26,32
10	Announcements	5	26,32
11	Timeline	5	26,32
12	Quiz	4	21,05
13	Profile	4	21,05
14	Lecturer's Profile	2	10,53
15	Join Course	2	10,53
16	Chatting	1	5,26
17	Feedback Form	1	5,26
18	Live Video Conference	1	5,26
19	Attendance List/Sheet	1	5,26
20	Online Storage	1	5,26

IOP Conf. Series: Materials Science and Engineering **325** (2018) 012011 doi:10.1088/1757-899X/325/1/012011

Of these features, participants once again being involved in the second workshop in which they start to design the high-fidelity prototype. In this session, they can decide how a certain feature will be displayed and behaved on the screen.

Here we had developed mobile-based learning management system that covers the features. The homepage of the application was figured out in figure 1. During the development process, we hired professional programmer to build our prototype on mobile platform. In this phase, we intensively communicate with the programmer to ensure that the application is match with the design that has been created by participants during workshops.

Figure 1. Screenshot of the application's homepage.

5. Discussion

After going through the workshops, this study accentuated three prominent points about participatory design in mobile application development.

Initially, regarding to the design process, there are two incompatibly characteristics of participants. One, participants are able to follow well the instruction given to them. They can do the task that requested by designer, such as, draw storyboard, making discussion and taking feedback, and build prototype. However, due to their background and experience in using the same application that will be built in this study, they are inclined to design an application that mimic well-known application in the field. In this part, participants are unable to explore and exploit their idea even further. This makes them lost their opportunity to obtain the most preferable application.

Moreover, mobile devices where the mobile application is running within has limited resources, especially in screen size, battery life, and storage size. Comparing to bigger devices such as laptop and PC, mobile devices has certain usage pattern either in positive and negative way. On the one hand, user can bring it anywhere anytime. On the other hand, user need to deal with the limitations. Participatory design has taken its part in this matter. With this approach, user or participant had an opportunity to determine their own usage pattern. For example, they can pick only important features in application's homepage as figured out in Table 1. The features that came out and listed in Table 1 was emerge from the user knowledge about the limitation of mobile devices and how to deal with it.

6. Concluding Remarks

Participatory design is a means for user to obtain preferable application. The main purpose of its usage is acquiring user requirements as match as possible with user needs. It is an opportunity for the user to determine how their application will behave. The goal of this approach is making a highly usable application. However, there are still challenges of making participant become actively express their idea so that the features of the application can perfectly match with their needs. Many research has been done to implement participatory design in various context. This research is among of them.

In this research, participatory design is being used in the context of mobile application development. The method helps user to explore their notion about the appearance, behavior, and features of mobile application based on their knowledge and experience. It can be done because their ideas are accommodated in several workshop sessions. In spite of that, there are still challenges gaining attention from participants during the whole process.

To sum up, this study has shown that participatory design could be an alternative approach in developing mobile application. It can accommodate the behavior of user in dealing with mobile device characteristics.

Further study can be done in developing and implementing working application that came from participatory design process in various context. Moreover, there are an opportunity also to do an evaluation study or user acceptance test regarding to applications generated from participatory design process.

References

[1] Statista 2016 *Number of smartphone users worldwide from 2014 to 2020 (in billions)* [Online] Available: https://www.statista.com/statistics/330695/number-of-smartphone-users-worldwide/ [Accessed: 28-Dec-2016].

[2] www.appbrain.com 2016 *Number of Android applications* [Online Available: https://www.appbrain.com/stats/number-of-android-apps [Accessed: 20-Jan-2017].

[3] Zhou W, Zhou Y, Jiang X, and Ning P 2012 Detecting repackaged smartphone applications in third-party android marketplaces *Proc. Second ACM Conf. Data Appl. Secur. Priv. (CODASKY '12)* pp 317–26.

[4] König-Ries B 2009 Challenges in Mobile Application Development *it - Inf. Technol.* **51(2)** pp 69–71.

[5] Abrahamsson P et al. 2004 Mobile-D: An Agile Approach for Mobile Application Development *Int. J. Serv. Ind. Manag.* pp 174–5.

[6] Lindberg S 2013 Participatory Design Workshops with Children with Cancer: Lessons Learned *Proc. 12th Int. Conf. Interact. Des. Child.* pp 332–5.

[7] Sanders E B-N and Stappers P J 2008 Co-creation and the new landscapes of design *CoDesign* **4(1)**, pp. 5–18.

[8] Harpe R D 2012 Lessons Learnt from The Participatory Design of A Mobile Care Data Application in A Resource-Restricted Context *Proc. of the 12th Participatory Design Conf. on Exploratory Papers Workshop Descriptions Industry Cases (PDC '12)* **2** p 125.

[9] Woźniak P, Romanowski A, Yantaç A E, and Fjeld M 2014 Notes from The Front Lines: Lessons Learnt from Designing for Improving Medical Imaging Data Sharing *Proc. 8th Nord. Conf. Human-Computer Interact. Fun, Fast, Found. (Nord. '14)* pp 381–90.

[10] Kautz K 2014 Participatory Design Activities and Agile Software Development *Hum. Benefit Through Diffus. Inf. Syst. Des. Sci. Res.* pp 303–16.

[11] Khaled R and Vasalou A 2014 Bridging Serious Games and Participatory Design *Int. J. Child-Computer Interact.* **2(2)** pp 93–100.

Vehicle Routing Problem Using Genetic Algorithm with Multi Compartment on Vegetable Distribution

Hari Kurnia [1], Elyza Gustri Wahyuni[2], Elang Cergas Pembrani[3], Syifa Tri Gardini [4], Silfa Kurnia Aditya[5]

[12345]Jl. Kaliurang KM 14,5 Sleman, DIY Tlp. (0274)895287, Fax (0274)895007 Yogyakarta 55584

14523317@students.uii.ac.id, elyza@uii.ac.id, 14523290@students.uii.ac.id, 14523232@students.uii.ac.id, 14523319@students.uii.ac.id

Abstract. The problem that is often gained by the industries of managing and distributing vegetables is how to distribute vegetables so that the quality of the vegetables can be maintained properly. The problems encountered include optimal route selection and little travel time or so-called TSP (Traveling Salesman Problem). These problems can be modeled using the Vehicle Routing Problem (VRP) algorithm with rating ranking, a cross order based crossing, and also order based mutation mutations on selected chromosomes. This study uses limitations using only 20 market points, 2 point warehouse (multi compartment) and 5 vehicles. It is determined that for one distribution, one vehicle can only distribute to 4 market points only from 1 particular warehouse, and also one such vehicle can only accommodate 100 kg capacity.

1. Introduction

Almost all Indonesians (97.29 percent) consume vegetables [1]. To comply the needs of vegetable consumption, in addition to the required quality improvement. Consumers love the vegetables in a fresh and hygienic state, while itself and seasonally to damage (easily rotten). Vegetables are also usually imported from other places far from the perimeter. Vegetables must be consumed as soon as possible due to limited time before decay or decreased quality.

Given the facts about vegetables described above, vegetable suppliers have their own challenges in distributing their vegetable products to the rest of the country. The challenge is how to choose an effective travel route. Effectively referred to here is the selection of travel routes with short distances. In addition, if the supplier has more than one supply warehouse, then another constraint will arise. This causes the supplier to be able to determine from what warehouse and where the purpose of the distribution is more efficient by considering the different distance between one a warehouse with another warehouse to a particular consumer.

This research is a development of 2 (two) previous research that has been done. The first study conducted by (Slamet et al, 2014) is the vehicle routing problem (VRP) with the genetic algorithm on the distribution of highland vegetables. The study has only one supplier or one warehouse alone as the origin of the vegetable distribution route. The second study conducted by (Tanujaya et al, 2013) is the application of genetic algorithms for vehicle routing problem-solving in PT.MIF. This study determines the parameters in GA is stimulated not separately and does not use GA optimization so that the resulting output is not good, fast and optimal.

Therefore, the authors conducted research on Vehicle Routing Problem (VRP) with the settlement using genetic algorithms that implement multi-compartment on vegetable distribution. This study aims to distribute vegetables to various regions in Java Island by considering the optimal mileage. The results of this study will be used for industries that want to supply vegetables to urban areas and also remote areas to minimize the occurrence of damage (rotten or not fresh) that occurs in the vegetables it carries.

1.1. Genetic Algorithm (GA)

On the search for solutions that sometimes require complex mathematical formulations to provide a definite solution [4].

To overcome this then used the calculation by a heuristic method that called genetic algorithm. Genetic algorithm is a search method based on the mechanism of natural selection and genetics [5]. In a genetic algorithm to produce an optimal solution, the search process is performed among a number of alternative optimal points based on a probabilistic function [6].

1.2. Vehicle Routing Problem (VRP)

VRP can be defined as the problem of designing a shipping vehicle path known to its capacity, operating from one depot to provide supply to a set of customers with known location and demand with one or more definite commodities [2].

VRP belongs to the Nondeterministic Polynomial-Hard (NPH) class which means that the parameters that influence are very complex and the solution can not be solved by a linear algorithm only because it takes a very long time, which generally uses a heuristic approach to find the solution [2].

Just like TSP, on VRP each vehicle must depart from a depot with a specific route to meet the demand nodes in the route and return to the original depot [3]. Here is a simple overview of the VRP using 2 vehicles, there are 1 depot and 12 nodes after the solve two routes with the minimum total distance.

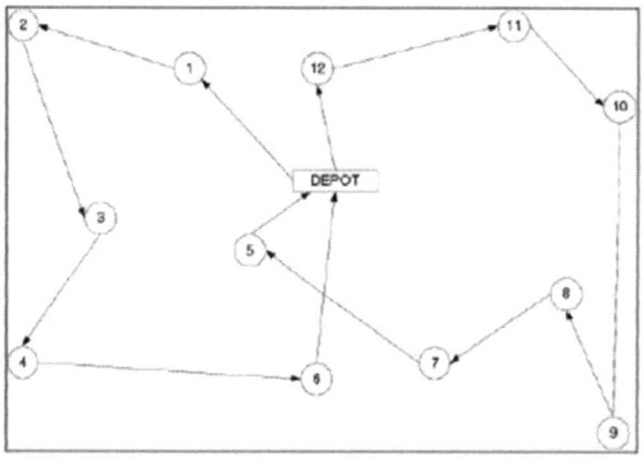

Figure 1. Simple VRP with number of vehicle 2

Each vehicle has a certain maximum operating time at which time is the maximum time for a vehicle to return to the depot [3].

2. RESEARCH METHOD

In this research method, will be discussed about research data, and also the formulation of a genetic algorithm to solve Vehicle Routing Problem (VRP).

2.1. Research Data

The first thing to do is determine the markets that will be used as a point of distribution of vegetables, and also the point of a warehouse that will be used. In this research will be used 20 point market for distribution and also 2 warehouses for vegetable storage. Then after determining the market points and also the point of the warehouse, the researcher takes the data of the distance between each market, the distance from each market to the warehouse, and also the warehouse distance to the warehouse. Distance value data can be obtained from google maps with units of km (kilometers). After the distance data was obtained, the researchers began compiling the systematical distribution of vegetables from 2 warehouse points to 20 market points. The data used for vegetable distribution systematics is the dummy data obtained from the random function in the excel software.

2.2. Initial Modelling
In the initial modeling of vehicle routing problem solving using a genetic algorithm, the following are formulated:

2.2.1. Identify Constrains.
In the vehicle routing problem, there are some limitations that must be fulfilled so that a solution can be considered as a viable solution. The boundaries that cause a solution to become unfeasible are called constraints. In order for a genetic algorithm to be executed to produce a viable solution, it is necessary to identify what factors make a solution improper.

2.2.2. Formulation of Objective Function and Fitness Function.
The objective function serves to measure the value of a solution. The result of this objective function is then converted into a fitness value that determines the quality of a solution by using the fitness function.

2.3. Model Optimization (Genetic Algorithm)
At this stage, model optimization is done to obtain optimal results. Optimization is done by changing parameters that determine how genetic algorithms work. The parameters in question are the initial population, population size, probability of crossing, and the probability of mutation. The genetic algorithm begins with the formation of the initial population of chromosomes formed by using random permutation algorithm. The next step in the genetic algorithm is the process of natural evolution that occurs continuously, namely selection, crossing, and mutation.

2.4. Ranking Selection
The first stage of natural evolution is selection, choosing which individuals will survive in the new population. The type of selection used in this study is rank selection, which is a stage of population selection that will live to the next generation by ordering the fitness value of each individual. Individuals to be selected as mothers are the individuals who have the highest fitness scores. Individuals who have the worst fitness score will die and be replaced by the crossed individuals.

2.5. Position Based Crossover
The second stage of natural evolution is a crossover, crossing the individual genes to produce new individuals who inherit the nature of the two parent. In this study, the type of crossing used is Position Based Crossover, which is a crossing stage applied to chromosomes that have the characteristics of permutation genes. The crossing is done by selecting several genes to cross, then the selected genes will be exchanged with their partner in the same positioning gene. Then another non-selected gene will adjust its sequence based on the same sequence of genes on the chromosome. In this study, the genes that will be selected for crossover are just destination genes only.

2.6. Binary Mutation dan Position Based Mutation
The third stage of natural evolution is the mutation, the sudden change of genes in the population and occurs randomly. In this study, the type of mutation used there are two, namely Binary Mutation and Position Based Mutation. Binary Mutation is a mutation stage used to convert a gene into another random gene. In this study, a binary mutation will be applied to the gene compartment. While Position Based Mutation is the stage applied to chromosomes that have the characteristics of permutation genes.

Mutations are made by exchanging selected gene positions to other random genes on the same chromosome. In this study, a position based mutation will be applied to the destination genes.

2.7. Model Analysis dan Evaluation
At this stage analysis and evaluation of the model applied to obtain the optimal model. If the model is still not optimal then it will be done changing the parameters of the genetic algorithm.

3. RESULT AND DISCUSSION
The results and discussion in this study include VRP modeling and also the process of formulating a genetic algorithm to solve Vehicle Routing Problem (VRP).

3.1. Research Data

3.1.1. Warehouse.
This research applies multi-compartment, that is source point that provides goods there is more than one point. Source point used in this study there are two, namely the warehouse located in (A) Boyolali and warehouse located in (B) Magelang.

3.1.2. Market.
Vegetables will be distributed to the markets spread in the area of Yogyakarta and surrounding areas. These markets will be the destination points in the model created. There are 20 destination points used in this research: (1) Pakem Market, (2) Prambanan Market, (3) Turi Market (Sleman), (4) Tempel Market, (5) Cebongan Market, (6) Colombo Market, (7) Kutu Market, (8) Condong Catur Market, (9) Srago Market, (10) Basin Market, (11) Pingit Market, (12) Klewer Market, (13) Jambon Market, (14) Muntilan Market, 15) PuluhWatu Market, (16) Godean Market, (17) Sentul Market Yogyakarta, (18) Ngasem Market, (19) Demangan Market, and (20) Jangkang Market.

3.1.3. Demand.
The amount of vegetable demand from the market will be considered to be the number of vegetables to be distributed to the destination points. In this study, all markets were uninformed and had the same amount of demand, 25 kg of vegetables for each market. This number of requests is a fixed number of requests.

3.1.4. Distance.
In this study, the distance to be taken from one point to another is obtained from the Google Maps map application. If there are several alternative paths to take, the path that occupies the top position in Google Maps. From here formed distance matrix that maps the distance of each point to another point.

3.1.5. Vehicle.
In this study, all vehicles have the same type and the same capacity, which is 100 kg per vehicle. The number of vehicles available for distribution of goods there are 5 vehicles, with the total capacity of the entire fleet is 500 kg of vegetables.

3.2. VRP Modelling
The VRP model created the model of transportation and distribution from the source points (warehouses) to the point of destination (the market) and back to the source point. In order to model VRP, it is necessary to identify constraints and problem limits and formulate objective functions and fitness functions.

3.3. Identify Constraints and Limitations
The VRP model created several constraints that must be a solution for a viable solution. There are some of the constraints and limitations applied to this research, among others ar each destination point is

visited exactly once, each vehicle departs from one of the specified source points, no vehicles distribute the goods beyond their capacity.

3.4. Formulation of Objective Functions and Fitness Functions

The objective function is a function used to measure the value of a solution. In this study, the objective function aims to calculate the total distance traveled by all vehicles. Here is the formula for calculating the total distance traveled by all vehicles in one individual problem solving:

$$sum\ d = \sum_{i=1}^{5}\left(\sum_{j=1}^{4} d_{ij} + d_{i0}\right) \quad \ldots\ldots\ldots\ldots\ldots\ldots\ldots\ldots\ldots (1)$$

The above formula, sum d means that the total distance is the number of the mileage of the i vehicle, and passes through four destination points (according to the capacity of the vehicle), then back again to the source point.

While the fitness function is used to calculate fitness that is the quality value of an individual problem-solving. Because the optimization will be done is minimization optimization (the smaller the distance the better), then the fitness function applied will change the small value into a great value fitness. Therefore, the fitness function uses a fairly simple formula:

$$f(d) = \frac{1}{d} \quad \ldots\ldots\ldots\ldots\ldots\ldots\ldots\ldots\ldots (2)$$

With (d) is the distance from the vehicle.

In this case, the number of possible solutions (including infeasible solution) in the search space (Nt) for this VRP problem is by looking at the genes that are on the chromosome. A chromosome has 25 genes which is divided into 5 genes representing the warehouses and 20 genes representing the markets. To calculate all possible solutions can uses the formula as below:

$$N_t = w^n * v! \quad \ldots\ldots\ldots\ldots\ldots\ldots\ldots\ldots\ldots (3)$$

With (w) is a number of warehouse, n is a number of genes which represents warehouse and v is a number of visited points.

$$N_s = i * k \quad \ldots\ldots\ldots\ldots\ldots\ldots\ldots\ldots\ldots (4)$$

Ns is the number of solutions generated from the calculation of the program to be run. With i is the number of iterations representing the number of generations to be generated and k is total solution per generation. The percentage of prospective search for a solution by genetic algorithm in the search space (Psearch) uses the formula as below:

$$P_{search} = \frac{N_s}{N_t} * 100\% \quad \ldots\ldots\ldots\ldots\ldots\ldots\ldots\ldots\ldots (5)$$

3.5. VRP Completion with Genetic Algorithm

The genetic algorithm is a searching technique and optimization technique based on evolutionary processes and changes in the genetic structure of living things [7]. In nature, chromosomes store the genetic information of an individual and consist of a set of genes. The characteristics and characteristics of an individual are determined by the genes, and these genes will be inherited to their offspring when breeding occurs. Individuals with good genes will survive over time, and individuals with bad genes will die. That means that the individual carrying the best genes and chromosomes will live and pass on the good characteristics of the offspring so that the average characteristic value of a population will continue to increase. Departing from this concept, genetic algorithms were developed with the aim of finding the optimal solution by crossing out existing solutions. Processes that occur in genetic algorithms can generally be seen in the following flow diagrams:

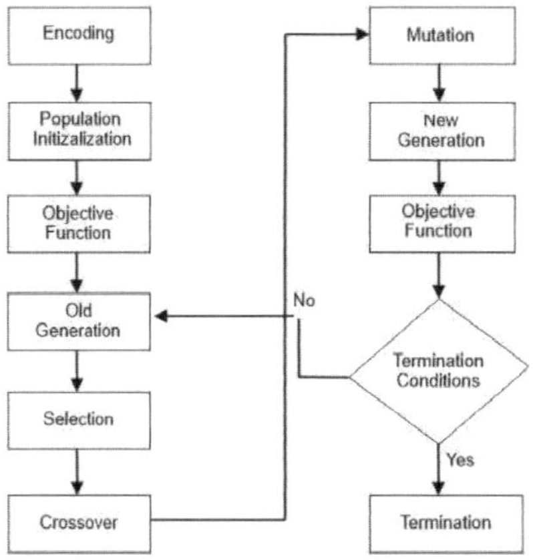

Figure 2. Genetic Algorithm Flow Diagram

The genetic algorithm consists of several stages, starting from the encoding stage, initial population initialization, selection, crossing, and mutation. After the algorithm is complete, it will also do the process of translation of the solution (decoding). Selection is used to determine which individual will survive, the cross producing new individuals that will replace the dead individuals so that the size of the population remains, and the mutation allows the emergence of new individuals unexpectedly. After the solution search is completed, then do the decoding, which translates the solution that has been coded into a form that can be understood by humans as the answer to the problems encountered.

3.6. Chromosome Representation

In the genetic algorithm, each individual represents every possible solution of the problem at hand. The solutions brought by each individual are encoded in the form of chromosomes, consisting of genes. In this study, the chromosome representation used is a chromosome representation of strings. In this chromosome, the source point is represented by a gene containing the letters of the alphabet, and the destination point is denoted by a gene containing numbers. The sequence and location of the source gene and the destination gene determine the origin and the points to be visited by each vehicle.

Table 1. Chromosome Representation

Vehicle. 1					Vehicle. 2					Vehicle. 3					Vehicle. 4					Vehicle. 5				
A	1	2	3	4	B	5	6	7	8	A	9	10	11	12	B	13	14	15	16	A	17	18	19	20

From the figure above, it can be seen that vehicle 3 travels from the source point A, then travels to the 9-10-11-12 destination and returns from the destination point 12 to the source point (A).

3.7. Selection

Selection technique used in this research is ranking selection technique, that is selection done by sorting population based on its fitness value. The highest fitness chromosome will rank first, and so on. Selection is chosen to reduce the possibility of an individual with a very large fitness dominating the population. By using the rankings, then the value of fitness is too large will not greatly affect the selection process is done randomly.

3.8. Crossover

The process of crossing is the process by which the chromosomes between two individuals are crossed to produce new individuals. The crossing process of two mains will result in two new individual children. Crosses aim to produce a new population with a higher average fitness value compared to the previous population.

Some individuals from the selection process will be randomly selected to cross each other. The selection is done randomly with a predetermined probability, is the probability of crossover (Pc). The expected frequency of the number of individuals to be crossed is Pc x population size. The crossover process used is a position based crossover where the position of the selected genes on one parent will be exchanged with selected genes on the other parent, and then another gene will adjust the sequence. In this study, the genes chosen to be crossed are only destination point genes, because if crossing involves source point genes the number of vehicles will not be regular.

Table 2. Parent 1 & Parent 2

Vehicle. 1					Vehicle. 2					Vehicle. 3					Vehicle. 4					Vehicle. 5				
B	5	18	2	17	B	10	20	1	9	A	12	14	16	6	A	4	8	7	19	B	15	13	11	3
B	14	13	6	19	B	10	8	16	9	B	3	11	7	20	B	17	1	5	18	B	15	4	2	12

Table 3. Descendant 1 & Descendant 2 :

Vehicle. 1					Vehicle. 2					Vehicle. 3					Vehicle. 4					Vehicle. 5				
B	5	2	6	17	A	10	20	1	9	B	12	14	16	4	B	8	7	19	18	B	15	13	11	3
B	14	13	2	6	A	10	8	16	9	B	3	11	7	20	B	17	1	5	19	A	18	15	4	12

3.9. Mutation

Of the population of selection and crossing, then carried out the process of mutation, which is a process of randomly converting genes. The selection of a mutated gene is randomly assigned to a predetermined probability, the mutation probability (Pm). The mutation process used is binary mutation and position based mutation.

Binary mutation is used in the source point gene, where the selected gene is replaced by another gene. In this study, binary mutation of the source point gene is done by converting the gene into (A) if previously (B), and vice versa.

Position based mutation is used in the destination point gene, where the gene that has been selected for the mutation will be exchanged for its position with another random destination point gene on the same chromosome.

3.10. Evaluation

Resolving the VRP problem resulted in the distribution of vegetable distribution routes from 2 warehouses to 20 markets with 5 vehicles. Details of the paths taken by each vehicle can be seen in Table 1. The algorithm is run with the following parameters: Population size is 100, Max generation is 1000, Probability of crossoveris 25%, Probability of mutation is 1%.

These parameters is determined by experiment. Population size of 40 is determined by having a look at the number of gen in a chromosome. Having a too small size of a population will risk the occurrence of a convergence (A population heavily dominated by a single chromosome). Max generation of 1000 is to ensure that the algorithm will yield a near-to-optimal solution, given enough iteration (generation). Crossover probability of 25% is determined by limiting the amount of offspring (new solutions) to a quarter of the population, so that the best parents would not be replaced by its less optimal offspring. As for the probability of mutation, probability of 1% is enough mutation (sudden change of a chromosome/solution), as higher chance would make it a different kind of genetic algorithm approach.

Table 4. The best solution of genetic algorithm

Number	Capacity (kg)	Visited Point		Distance Estimation
1	100	1. Magelang Warehouse	4. Sentul Market Yogyakarta	93.6 kilometers

		2. Jangkang Market 3. Pakem Market	5. Ngasem Market 6. Magelang Warehouse	
2	100	1. Magelang Warehouse 2. Muntilan Market 3. Condong Catur Market	4. Demangan Market 5. Kutu Market 6. Magelang Warehouse	94.8 kilometers
3	100	1. Magelang Warehouse 2. Turi (Sleman) Market 3. Godean Market	4. Pingit Market 5. Tempel Market 6. Magelang Warehouse	93.7 kilometers
4	100	1. Boyolali Warehouse 2. PuluhWatu Market 3. Basin Market	4. Klewer Market 5. Srago Market 6. Boyolali Warehouse	83.9 kilometers
5	100	1. Magelang Warehouse 2. Colombo Market 3. Jambon Market	4. Prambanan Market 5. Cebongan Market 6. Magelang Warehouse	122.2 kilometers

The best solution with these parameters resulted in a total distance of about 488.2 km and a fitness value of 0.00205, quite different when compared to the best solution in the first generation with a total distance of about 675.8 km and a fitness value of 0.0015.

The best solution the genetic algorithm came up with is a result of both the crossover method and the mutation method. By using position based crossover, the algorithm will search for a better path for each of the vehicle included. The crossover only affects gens that represents destination nodes, so that it won't mess with the amount of load a single vehicle could carry, in case each vehicle has different capacity. On the other hand, the mutation will be able to change both gens that represents source nodes (compartments) and gens that represents destination nodes. The gens that represents source nodes will mutate by means of binary mutation, and the gens that represents destination nodes will mutate by means of position based mutation. This way, the algorithm will be able to find the optimal solution within the feasible solution space and won't produce too much solutions that are infeasible.

The number of possible solutions (including infeasible solution) in the search space (Nt) for this VRP problem is about $2^5 * 20\,!$. While the number of candidate solutions that have been evaluated by the genetic algorithm exists:

Ns = (1000 generations) x (40 solutions / generation)
= 40,000 potential solutions

The percentage of prospective search for a solution by genetic algorithm in the search space (Psearch) is:

Psearch = (Ns / Nt) x 100%
= $(40.000 / (2^5.20!))$ X 100%
= 5,1379 x 10-14%

From that percentage, it appears that the genetic algorithm only searches for solutions on a very small portion of the overall probability of existing solutions, and has gained fairly optimal results.

4. CONCLUSION

From the tests that have been done, the population size of 100, max 1000 generation, the probability of crossover 25%, and the probability of the 1% mutation the best solution resulted in a total distance of 488.2 km and total fitness of 0.00205 which differed considerably from the first generation is the total distance around 675,8 km and fitness value 0,0015.

The authors conclude that the genetic algorithm only searches for solutions in a very small part of the overall probability of the existing solution, and has gained fairly optimal results with the total distance obtained less than the total distance obtained in the first generation and get a fitness value that is not much different from the fitness value of the first generation.

Suggestions for this research in the future is to add the optimal route calculation of travel time taking into account the congestion factor. So that the distribution of vegetables will be more optimal.

REFERENCES

[1] Nasional H G 2017 *Konsumsi Buah Dan Sayur Susenas Maret 2016.*
[2] Hermansyah B, Sains F, Teknologi D A N, Islam U, Sultan N, and Kasim S 2011 *(VRP) Menggunakan Algoritma Genetika.*
[3] Tanujaya W, Retno D, Dewi S, and Endah D 1976 *Penerapan Algoritma Genetik Untuk Penyelesaian Masalah Vehicle Routing Di Pt . Mif* pp 92–102.
[4] Hannawati A 2002 *Pencarian Rute Optimum Menggunakan Algoritma Genetika* **2(2)** pp. 78–83.
[5] Paper C 2015 *Optimasi Fungsi Tak Berkendala Menggunakan Algoritma Genetika Terdistribusi dengan Pengkodean Real Optimasi Fungsi Tak Berkendala Menggunakan Algoritma Genetika Terdistribusi dengan Pengkodean Real.*
[6] Michalewicz Z et al 1996 Evolutionary algorithms for constrained engineering problems *Computers & Industrial Eng.* **30.4** pp 851-70.
[7] Slamet, Setiawan A, Siregar H H, and Kustiyo A 2014 Vehicle Routing Problem (VRP) dengan Algoritma Genetika pada Pendistribusian Sayuran Dataran Tinggi *Jurnal Teknologi Industri Pertanian* **24.1**.

IOP Conf. Series: Materials Science and Engineering **325** (2018) 012013 doi:10.1088/1757-899X/325/1/012013

The Technology of LiFi: A Brief Introduction

E Ramadhani [1], **G P Mahardika** [2]

[1,2] Department of Informatics, Universitas Islam Indonesia, Yogyakarta, Indonesia

Email: erika@uii.ac.id, galang.prihadi@uii.ac.id

Abstract. Light Fidelity (LiFi) is a Visible Light Communication (VLC) based technology that making a light as a media of communication replacing the cable wire communication. LiFi is evolve to overcome the rate speed in WiFi, while using LiFi the rate speed can reach until 14 Gbps. This paper presents an introduction of the LiFi technology including the architecture, modulation, performance, and the challenges. The result of this paper can be used as a reference and knowledge to develop some of the LiFi technology.

1. Introduction

LiFi known as light fidelity was introduced first time by Prof. Harald Haas on July 2011 at TED Global Talk. LiFi is based on Visual Light Communication (VLC) that using light emitting diodes (LEDs) to fully networked wireless system [1]. LiFi enables the electronic device to connect to the internet with no wire. In order to make a communication line between node, a LiFi will need a transceiver to transmit and receive the data. This transceiver will have a modulation technique to make the LED enable to carry the data using the light. The emergence of LiFi is to overcome the shortage of the current technology. We all know that right now WiFi is the most used technology to connect many devices to the internet. As time comes by, the use of internet based devices is increased. This increasing made the capacity of WiFi is reduced due the limitation of radio frequency resources.

According to [2], LiFi and WiFi has the difference related to the congestion, density, security, safety, and speed. The more WiFi enabled device is exist, the congestion may occur. In the technology of WiFi we can't add more routers if the user is increased, while we can add the light in LiFi. Efficiency and safety of the internet are the dominating issues right now. The performance of LiFi is claimed that more better than the performance of WiFi. The rate speed of LiFi is 1000 times faster than WiFi [2]. For safety of the internet, LiFi is more secure than the WiFi based on the spread of the signal. LiFi has a light characteristic that light cannot go through the wall. It is different from the signal of WiFi can go through anywhere. Based on those two technologies, in a simple conclusion is LiFi has more secure communication rather than the WiFi. The vulnerability exists if there are leakage in the wall while having an indoor communication. Security threat that may exist is an intruder can spoof the data using the leakage wall.

This paper provides an explanation of the newly developed technology that is LiFi. The paper will discuss about the architecture, modulation, performance, and the challenges. The structure of this paper is as follows: in the remaining part of this section we will discuss the LiFi concept based on VLC. In section 3 we discuss the architecture of LiFi. In section 4 we further analyze the performance of LiFi based on the rate speed and security comparing with the WiFi. In section 5 we discuss about the challenge in LiFi technology, it includes a future research based on the security issue. All the result of this paper is summarized in section 6. The result of this paper can be used as a reference and knowledge when developing a LiFi technology.

2. What is LiFi

According to figure 1, LiFi technology consist of LED Lamp as the media transmission and photo detector as a receiver of transmitted data. Lamp driver is needed to make LED working properly. While amplification and processing are responsible to manage the signal that comes from the photo detector.

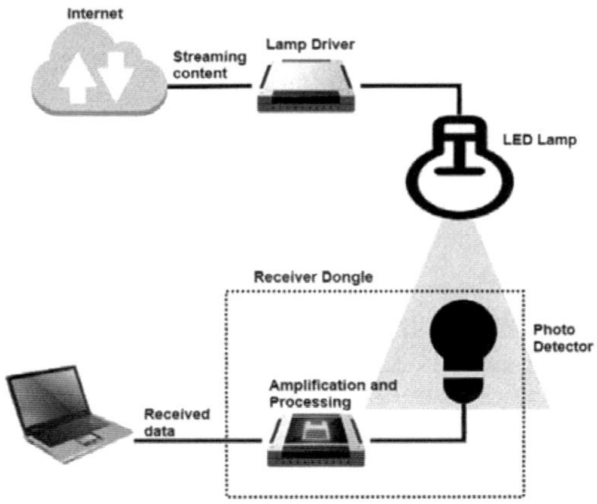

Figure 1. Basic Concept Diagram LiFi

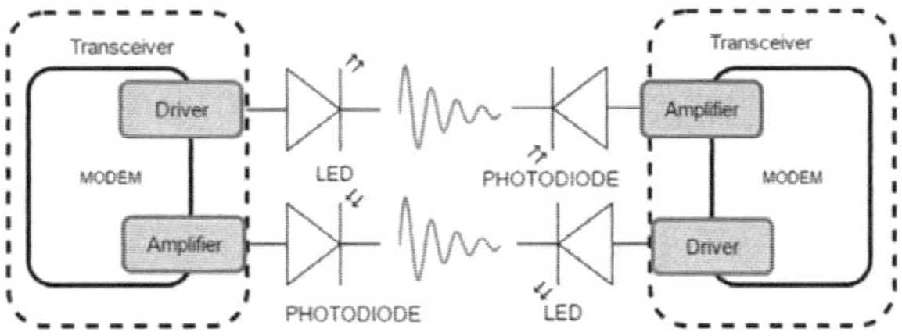

Figure 2. Transceiver LiFi based on VLC

Basic concept for working principle in LiFi Technology are pointing into: Transceiver and Light as a media transmission. Figure 2 is a basic concept block diagram for LiFi. This basic concept indicates as a duplex communication. The rates of LiFi is 14Gbps using three off- the- shelf laser diodes (red, green, and blue) and predict the rate until 100Gbps when the whole visible spectrum is used.

According to [3] LiFi and VLC used a similar medium as a data communication that is light. The difference between LiFi and VLC is VLC has a unidirectional, point-to-point light communication at low data rates. While the LiFi technology is fully networked, bidirectional, and high-speed wireless communication. Others said LiFi is the incorporation of WiFi and VLC[2].

2.1. Tansceiver

Transceiver is a block that act as a transmitter and receiver at the same time. This transceiver consists of LED to transmit the light and photodiode to receive the light. Amplifier is embedded to strength the power of light received from the photodiode. The modem is used to modulate and demodulate the signal. The signal that comes from the photodiode is analog and it converts into digital in the modem. While the signal that ready to transmit, the digital signal convert into analog signal in the modem and sent by LED. The driver before the LED operates to drive the current of the LED in order to get the flickering. The flickering is functioning the LED for data transmission, if LED is ON then it transmits digital '1' and if OFF, it transmits digital '0'[4][5].

2.2. Modulation

The modulation signals is used to switch LED at desired frequencies that contains information to be transmitted. According to [6] there are several technique of modulation in LiFi. Modulation techniques is needed in order the communication is still available even the illumination is not required. Because of that, a modulation technique may support a dimmable illumination. The variation in intensity of light corresponding to the information in the message signal. There are many typical of modulation in LiFi i.e. Single Carrier Modulation (SCM), Multiple Carrier Modulation (MCM), and Colour Modulation.

2.2.1. Single Carrier Modulation (SCM)

SCM is a modulation multiplexed any number of signal, then modulated individually with different frequency[7]. SCM using a microwave as a subcarrier and an optical carrier. Modulation technique in SCM are on-off keying (OOK), pulse-position modulation (PPM), and pulse-amplitude modulation (PAM). SCM is suitable for low-to- moderate data rates applications.

2.2.2. Multiple Carrier Modulation (MCM)

Because of the performance of SCM degrades as their spectral efficiency increase. SCM also require complex equalization process when employed at high data rates. MCM is developed to replace the disadvantage of SCM. There are several kinds of MCM i.e. OFDM. The difference between SCM and MCM summarised in table 1.

Table 1. Difference technique of LiFi Modulation

Modulation			**Characteristic**
SCM	OOK		Transmit data by sequentially turning on and off the LED, the reliable communication range would decrease at low dimming levels, and increasing and decreasing the brighteness of the LED would cause the data rate to decrease
	PPM		Provide efficient and dimming support in variable pulse position modulation (VPPM)
	PAM		Sensitive to signal distortion[8], combination with other modulation technique to get a better performance
MCM	OFDM		Excellent for situations where multiple transmitters are used simultaneously, avoid shadowing effects, the interference can be mitigated by shifting the system bandwidth to higher frequency[9]
		DCO-OFDM	The substantial energy dissipation due to the biasing [10]
		ACO-OFDM	Efficient in term of optical power for lower SNR value for IM/DD channel [11]
		PAM-DMT	Better optical power efficiency compared to DCO-OFDM [12]
		AHO-OFDM	Support various dimming targets to achieve system performance [12]
		Flip-OFDM	Equivalent with the ACO-OFDM in term of spectral efficiency and error performance[13]
		U-OFDM	Equivalent with the ACO-OFDM in term of spectral efficiency and error performance[13]

3. The Architecture of LiFi

According to [14] the paper categorized the architecture based on layered. Figure 3 is the layered architecture of LiFi. In layered architecture, LiFi consist of 3 stages i.e. application layer, MAC layer, and Physical Layer. IEEE 802.15.7 defines only two standard i.e. PHY and MAC layer.

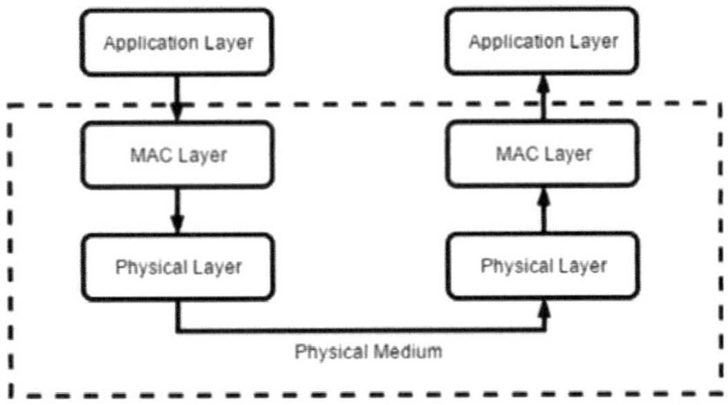

Figure 3. Layered architecture VLC [14]

3.1. IEEE 802.15.7 Layered Architecture

3.1.1. PHY Layer

PHY layer responsible in transmission and reception, activation and deactivation of optical transceiver, and detection of state of transmission channel, is it idle or busy state. There are 3 operation modes in PHY layer. The differences of each operation modes are show in table 2.

Table 2. Operation Modes PHY Layer [15]

Operation Modes	Usage	Categories	Rate
PHY I	Outdoor	Low	11.6 Kpbs – 266.6 Kbps
PHY II	Indoor	Moderate	1.25 Mbps – 96 Mbps
PHY III	Multiple optical transceiver	CSK Modulation	12 Mbps – 96 Mbps

3.1.2. MAC Layer

Three network topologies are defined in MAC layer: Peer to peer, star, and broadcast [14].

a) Peer to peer
   There are two device that communicate. One of them is act as a coordinator.
b) Star
   Communication happens in several devices. One of them is act as a coordinator and it's used as a illumination infrastructure.
c) Broadcast
   One device i.e . a coordinator sends data to a several devices. The communication is unidirectional way.

3.2. Propagation Channel

Propagation channel in LiFi is not different from VLC. According to [16] indoor environment characterized by six different link configuration refers to IR links. Transmitter and receiver communicate in two criterions, i.e. direct or indirect line-of-sight (LOS) that required in the propagation channel. These two criterions are based on degree of directionality of the transmitter and receiver (LOS) others is based on the reflection of the light (non-LOS). In the LOS the links between the transmitter and receiver is pointing or directed each other. While in the non-LOS the light is spreading by the reflection of the ciling or diffusely reflecting surface. The characteristic of each criterion summarized on table 3.

Table 3. Characteristic of IR Link

	Directed	**Hybrid**	**Nondirected**
Line of Sight	• Maximizes power efficiency • Minimize path loss and reception of ambient light noise	Combine transmitter and receiver having different degrees of direcionality	• Wide-angle transmitter and receivers • Convenient to use, for mobile terminals
Non Line of Sight	• Rely upon reflection of the light from ceiling or some other diffusely reflecting surface • Increase link robustness and ease of use • Allowing the link to operate even when barriers • Referred to as a Diffuse link		

The important parameter to get the high data rates is the availability of line-of-sight (LOS) optical link [17]. A non-directed LOS transmission is like to limit the achievable data rates. While the lighting scenario may vary, it is important to adaptation a dynamic rate to achieve a robust VLC link. According to [17] LOS is maybe no longer needed. K.D. Langer built rate-adaptive visible light communication at 500Mbps arrives at plug and play. He developed a bidirectional high-speed real time VLC system.

3.3. The topologies of LiFi

According to [2], the paper experiencing build a proof-of-concept prototype VLC HetNet. This prototype is a concept that using a diverse spectrum to provide high quality-of –service in indoor environment. There is an additional tier in the wireless HetNet using indoor gigabit small-cells to offer additional wireless capacity. The prof-of-concept experiment has two model such as hybrid model and aggregated model. In hybrid model, the system is used WiFi to connect to the internet but the downlink of a user is connected through a LiFi link. While the aggregated system is user connected on WiFi and LiFi in parallel. Unidirectional LiFi link is used in hybrid system to supplement the WiFi downlink. In aggregated system both WiFi and LiFi using bi-directional communication to improve the throughput and achieve a robust connectivity.

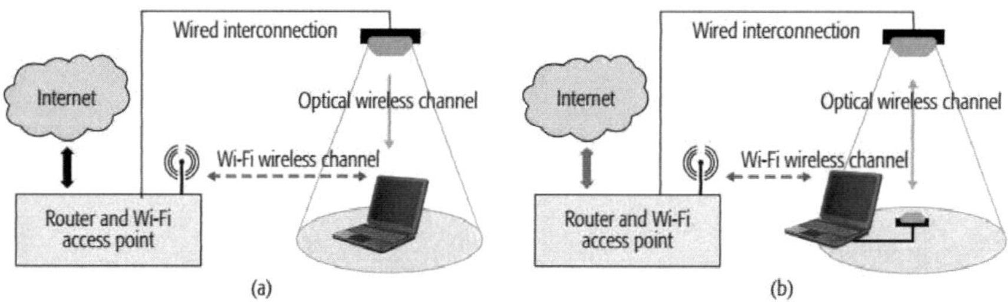

Figure 4. (a) Hybrid system and (b) aggregated system [2]

The goal of this concept is to achieve a good performance. In the hybrid system the throughput depends only on the capacity of the LiFi downlink while degrading as the distance increase. Therefore, in the aggregation system enhances the available reliable network communication. Hybrid system also can be a solution for LiFi channel blockage when implementing a mobile device.

4. Performance of LiFi

Efficiency and safety of the internet are the dominating issues now. Lifi was found in 2011 by Scientist Harold Haas from UK. The design is to overcome the disadvantage of WiFi. The speed of Wifi is up to 1500mbps and it's not sufficient to accommodate a huge user.

Table 4. Speed and Standards of WiFi

Standard	Release Date	Max Speed
802.11b	1999	11 Mbps
802.11a	1999	54 Mbps
802.11g	2002	54 Mbps
802.11n	2007	72-600 Mbps
802.11ac	2013	433 Mbps – 1.3 Gps

Table 5. Speed of LiFi

No	Reference	Modulation	Data Rate
1	[18]	OOK	803 Mbps
2	[19]	OFDM	2.1 Gbps
3	[20]	DMT	3.4 Gbps
4	[16]	PPM	30 Mbps
5	[21]	PAM	20Mbps
6	[22]	CAP	1.1 Gbps

LiFi enabling the system using fully networked wireless communication and could provide a connection that's 100 times faster than WiFi. It can reach speeds up to 3Gpbs by using DMT modulation [20]. The other speed of LiFi with the different modulation also shown in the table 5. M.D Renzo et all using Spatial Modulation for MIMO Wireless System in LiFi, the transmit speed is up to 10Gbps. The rate speed of LiFi can be higher than 3Gbps while the technology is on research and developing. The rate speed of LiFi is based on the using of the modulation. Dr. P. Kuppusamy et.all doing a survey of Lifi and comparing it with Wifi, there are several characteristics that used for comparing. The difference of LiFi and WiFi is showed in table 6.

For security, LiFi is more tought than a WiFi. It is because the coverage area of LiFi is only on their illuminate area. The signal of LiFi cannot go through the wall. While the signal of WiFi can go through the wall, it can cause the vulnerabilities in data loss and data leakage. According to [25] security issues in VLC focus on aspect of basic physical characteristics of the communication channel. The paper also analyzes the risk of signal jamming, data snooping, modification, and MAC-level security. The result is VLC infrastructure is particularly prone to data-security risk. According to [26] one of the security issue in LiFi is eavesdropping, an attack happen by obtaining the signal that come from the gap between the floor and door, crack inside the flooring or from partially shielded windows.

Table 6. The difference of LiFi and WiFi [23][5][24]

Parameter	LiFi	WiFi
Transmitter	LED	Antenna
Receiver	LED	Antenna
Inbuilt Device	Under research and development	WiFi Card/Chip
Average Operation Speed	Greater than 10Gbps (under research)	150-600 Mbps
Frequency band	1000 times of THz	2.4 GHz
Standard	IEEE 802.15.xx	IEEE 802.11xx
No of users	All over under the lamp.	Depend on access point.
Data Transmission	Bits	Radio waves
Coverage Area	10 meters	20 – 100 meters varies based on type of transmission power and antenna
Interference	No interference issues with RF waves	Interference with neighbor AP routers
Topology	Point to Point	Point to Multipoint
Communication	Based on Visible Light Communication	Based on Radio Frequency Communication
Efficiency	More, LEDs consume less energy and highly efficient	Less, Radio Base Stations consume high amount of energy
Availability	Anywhere, available in airplanes and underwater	Limited
Secure	More secure because light waves cannot penetrate through walls and cannot be intercept by anyone outside the illumination of LED	Less secure because of high penetrating power of radio waves, anyone can intercept
Network topology	Point-to-point	Point- to- multipoint
Suitability	Suitable for high data rates and secure communication	Suitable for Aps with high coverage regions
Signal-to-Noise Ratio	Very high	Maybe more
Power consumption	Less	More
Environment Impact	Low	Medium

5. The Challenges

The challenge in LiFi is based on point that has been discussed in literature and the infrastructure that already available. Including the real users of LiFi technology is how to response.

5.1. Modulation

The key of LiFi communication is the using of modulation. A modulation in LiFi is to carry a binary data by turning the LED on and off quickly. There are many aspects in LiFi related to modulation, illumination and dimming scheme is the first concern. Illumination is the spread of the light that making the LEDs can be as a media in data communication. The challenge is how the modulation is enabling the illumination of LEDs in order can send the data while the illumination is low. While the dimming process is to proportional of LEDs brightness. The challenge in dimming technology is how the LiFi can fulfill the user satisfaction in order the dimming of LED can stay safe for the user [27].

5.2. Interference

In optical illumination based on data communication, the hard part is to provide the optical uplink service. It is because the uplink service can interfere the downlink signal. These problem is one of the challenge in the interference signal issue. In LiFi, the transmitter should be able to maintain a directional link during the transmission[27][28].

5.3. Infrastucture

The basic infrastructure in LiFi is indoor and outdoor. Same as in the optical characteristic, a LiFi also has an effect shadowing while transmission. This shadowing effect off course will give an effect in the process of sending and receiving the data. There is a few research about the effect of shadowing in the LiFi communication [29].

5.4. Security

According to [26] a threat like eavesdropping can happen in LiFi. It happens when there is gap between the floor and the door, the light may spread out between them. The crack from inside floor and shielded windows also can be a leakage.

5.5. Coverage

LiFi is a technology that has a good perform in an indoor infrastructure while it's not happen in the outdoor area. The coverage in outdoor area for LiFi needs to be set up in order the quality of connection can give a good performance. According to [2], LiFi is integrated with the WiFi to get a good performance in an outdoor or in a mobile infrastructure.

6. Summary and Future Scope

In this paper, we outlined the working of LiFi, the modulation that its used, the architecture, the performance, and at last the challenges. The purpose of LiFi technology is to provide a high-speed data communication using visible light spectrum. Now LiFi is on-going of research, it has a potential advantage that can make a supplement RF communication and can be used to improve wireless network performance. Although LiFi has a good performance in the transfer rate, LiFi is not good enough when deploy in an outdoor in sunlight or other condition. LiFi will probably not completely replace WiFi, these two technologies can be used together to achieve more efficient and secure network.

Paris is one place that has been implemented the LiFi technology. There are offices, hospitals, retail stores, public street lighting as well as metro station at L Defense, and Curtius museum in Liege in 2012. The four key criteria that is used are: the LiFi can operate within a complex physical environment, the system can provide the bi-directional communication, traffic and security information can be pushed to smartphones, system of audio-messaging is developed to enable tourist and people to be guided.

In future, the work can be done for analyzing the security threats based on the architecture, communication, and implementation of LiFi according to CIA triad concept. It is also can be done by doing research based on the challenge in the LiFi technology.

References

[1] Haas H, Yin L, Wang Y, and Chen C 2016 What is LiFi? *J. Light. Technol.* **34** pp 1533–44.
[2] Ayyash M *et al* 2016 Coexistence of WiFi and LiFi toward 5G: Concepts, opportunities, and challenges *IEEE Commun. Mag.* **54(2)** pp 64–71.
[3] pureLiFi 2017 Shedding Light on LiFi *PureLiFi*.
[4] Singh P Y P and Bijnor U P 2013 A comparative and critical technical study of the Li-Fi *(A Future Communication) V / S Wi-Fi* **2(4)** pp 2011–3.
[5] Miteku N 2015 Li-Fi over Wi-Fi in internet data communication *Ijireeice* **3(12)** pp 153–9.
[6] Islim M S and Haas H 2016 Modulation techniques for Li-Fi Mohamed *ZTE Commun.* **14** pp 29–40.
[7] Revathi S and Aarthi G Performance analysis of Wave Length Division and Sub Carrier Multiplexing using different modulation techniques *Int. J. Eng. Res. Appl.* **1(2)** pp 317–20.
[8] Ghassemlooy Z, Yuan Y, Tang X, and Luo P *SVM Detection for Superposed Pulse Amplitude Modulation in Visible Light Communications*.
[9] Afgani M, Haas H, Elgala H, and Knipp D 2006 Visible light communication using OFDM *Proc. 2nd Int. Conf. Testbeds Res. Infrastructures Dev. Networks Communities* pp 129–34,.
[10] Islim M S, Tsonev D, and Haas H *A Generalized Solution to the Spectral Efficiency Loss in Unipolar Optical OFDM-based Systems*.
[11] Dissanayake S D, Member S, Armstrong J, and Member S 2013 Comparison of ACO-OFDM, DCO-OFDM and ADO-OFDM *IM / DD Systems* **31(7)** pp 1063–72.
[12] Ganesan S, Mesleh R, and Haas H 2014 *On the Performance of Spatial Modulation OFDM* pp 1–5.
[13] Fernando N, Hong Y, and Viterbo E 2011 *Flip-OFDM for Unipolar Communication Systems*.
[14] Khan L U 2016 Visible light communication: Applications, architecture, standardization and research challenges *Digit. Commun. Networks* **3(2)** pp 78–88.
[15] Ley-Bosch C, Alonso-González I, Sánchez-Rodríguez D, and Ramírez-Casañas C 2016 Evaluation of the effects of hidden node problems in IEEE 802.15.7 uplink performance *Sensors* **16(2)** p 216.
[16] Kahn J M and Barry J R 1997 *Wireless Infrared Communications* **9219** no. 97.
[17] Langer K, Hilt J, Schulz D, Hartlieb F, Kottke C, and Grobe L 2015 *Rate-adaptive visible light communication at 500Mb / s arrives at plug and play* no. November 2013.
[18] Vucic J *et al* 2010 803 Mbit/s Visible Light WDM Link based on DMT Modulation of a Single RGB LED Luminary *Opt. Fiber Commun. (OFC), collocated Natl. Fiber Opt. Eng. Conf. 2010 Conf.* no. April pp 7–9.
[19] Cossu G, Khalid A M, Choudhury P, Corsini R, and Ciaramella E 2012 2.1 Gbit/s Visible Optical Wireless Transmission *European Conf. and Exhibition on Optical Communication* p. P4.16.
[20] Cossu G, Khalid A M, Choudhury P, Corsini R, and Ciaramella E 2012 34 Gbit/s visible optical wireless transmission based on RGB LED *Opt. Express* **20(26)** p B501.
[21] Li J F, Huang Z T, Zhang R Q, Zeng F X, Jiang M, and Ji Y F 2013 Superposed pulse amplitude modulation for visible light communication *Opt Express* **21(25)** p. 31006–11.
[22] Wu F, Lin C, Wei C, and Chen C 2012 1.1-Gb/s White-LED-based visible light communication employing carrier-less amplitude and phase modulation *IEEE Photonics Technol. Lett.* **24(19)** pp 1730–2.
[23] Kuppusamy P, Muthuraj S, and Gopinath S 2016 Survey and challenges of Li-Fi with comparison of Wi-Fi *Proc. 2016 IEEE Int. Conf. Wirel. Commun. Signal Process. Networking (WiSPNET 2016)* no. March pp 896–9.
[24] Virk G K and Fidelity L 2015 Li-Fi : A new communication mechanism *Int. J. Comput. Appl.*

118(15) pp 1–4.
[25] Blinowski G 2015 Security issues in visible light communication systems *IFAC-PapersOnLine* **28(4)** pp 234–9.
[26] Classen J, Chen J, Steinmetzer D, Hollick M, and Knightly E *The Spy Next Door: Eavesdropping on High Throughput Visible Light Communications*.
[27] Bao X, Yu G, Dai J, and Zhu X 2015 Li-Fi: Light fidelity-a survey *Wirel. Networks* **21(6)** pp 1879–89.
[28] Lee S J and Jung S Y 2012 A SNR analysis of the visible light channel environment for visible light communication *APCC 2012 - 18th Asia-Pacific Conf. Commun. "Green Smart Commun. IT Innov."* pp 709–12.
[29] Khandal D and Sakshi J 2014 Li-Fi (Light Fidelity) - The future technology in wireless communication *Int. J. Inf. Comput. Technol.* **4(16)** pp 1687–94.

IOP Conf. Series: Materials Science and Engineering **325** (2018) 012014 doi:10.1088/1757-899X/325/1/012014

Numerical Simulation of Blood Flow in Human Artery Using (A, Q) and (A, u) Systems

Sudi Mungkasi and Inge Wijayanti Budiawan

Department of Mathematics, Faculty of Science and Technology,
Sanata Dharma University, Mrican, Tromol Pos 29, Yogyakarta 55002, Indonesia

E-mail: sudi@usd.ac.id

Abstract. In this paper, we model blood flow in human artery in the form of (A, Q) and (A, u) systems, then we use the Lax-Friedrichs finite volume method to find the numerical solution of each model. Here A represents the cross sectional area of the artery, Q denotes the discharge of the blood flow, and u is the velocity of the blood flow. We simulate the numerical scheme of each model and investigate how the blood pressure pulse propagates in human artery. Particularly, we use the residual of A to determine which system is better numerically. We obtain that the (A, Q) system is better numerically than the (A, u) system, because the absolute of the residual of A using the (A, Q) system is smaller than the absolute of the residual of A using the (A, u) system.

1. Introduction

Applications of Mathematics have been growing in various fields. Mathematics can be applied in various problems [3, 4], such as blood flow problems [1]. Modelling and simulation of blood flow are useful for human life.

There are a number of mathematical models for blood flow [2, 6]. Two of them are blood flow models in the form of (A, Q) and (A, u) systems [6]. Here A represents the cross sectional area of the artery, Q denotes the discharge of the blood flow, and u is the velocity of the blood flow. These (A, Q) and (A, u) systems can be used to explain phenomena of blood flow in human artery.

In this paper, we focus on modelling blood flow in human artery in the form of (A, Q) and (A, u) systems, solving both systems numerically, simulating both numerical solutions, and determining which system is better numerically. We use a finite volume method, because it is able to solve both continuous and discontinuous problems. Note that blood flow model has either continuous or discontinuous solution even though the initial condition is continuous. Then, we use the residual of A of each system to determine which system is better numerically.

This paper is organised as follows. We provide mathematical models and numerical methods for blood flow in Section 2. Research results and discussion are given in Section 3. We conclude the paper with some remarks in Section 4.

2. Models and methods

In this section, we model blood flow in human artery in the form of (A, Q) and (A, u) systems with several assumptions, and find the numerical method for each model. Blood flow models are derived from mass and momentum conservation laws [6].

We illustrate the shape of human artery in Figure 1. We denote the time variable t and the one-dimensional space variable z. We define that at an arbitrary cross section $S(z, t)$ of the artery

$$A(z,t) = \int_S d\sigma, \tag{2.1}$$

$$u(z,t) = \frac{1}{A}\int_S \hat{u}\, d\sigma, \tag{2.2}$$

and

$$p(z,t) = \frac{1}{A}\int_S \hat{p}\, d\sigma, \tag{2.3}$$

where A is the cross section area, u is the average velocity, and p is the average pressure, $\hat{u}(z,\sigma,t)$ is the velocity, $\hat{p}(z,\sigma,t)$ is the pressure. We define $Q(z,t) = A(z,t)u(z,t)$ as the volume flux at $S(z,t)$. We assume that blood is incompressible fluid, so the viscosity and the density of blood are constant.

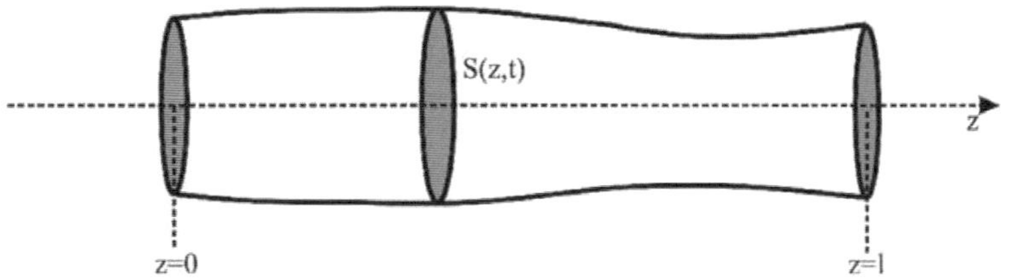

Figure 1. Illustration of human artery.

We have assumed that the shape of human artery is a cylindrical tube with circular cross section $S(z,t)$, and the circular cross section area does not depend on space variables x and y, but depends on the space variable z as well as the time variable t.

2.1. Mass conservation

Mass conservation law states that mass is neither created nor destroyed, so the total of the nett mass flux flowing out of a control volume and the rate of change of mass within a control volume equals to zero. We can write this statement as an equation

$$\rho\frac{dV}{dt} + \rho Q(l,t) - \rho Q(0,t) = 0, \tag{2.1.1}$$

where ρ is the density of blood and V is defined as follows

$$V(t) = \int_0^l A(z,t)\, dz. \tag{2.1.2}$$

Then, we can rewrite equation (2.1.1) as follows

$$\rho\frac{d}{dt}\int_0^l A(z,t)\, dz + \rho\int_0^l \frac{\partial Q}{\partial z}\, dz = 0. \tag{2.1.3}$$

Because the length l of the control volume is constant, we can rewrite equation (2.1.3) as follows

$$\rho\int_0^l \left(\frac{\partial A}{\partial z} + \frac{\partial Q}{\partial z}\right) dz = 0. \tag{2.1.4}$$

Equation (2.1.4) is satisfied for any value of l, so we obtain the mass conservation equation

$$\frac{\partial A}{\partial z} + \frac{\partial Q}{\partial z} = 0. \tag{2.1.5}$$

2.2. Momentum conservation

The Newton's second law states that the change of momentum within a system equals to the total of the applied forces. We assume that there is no flux through the artery walls, so the total of the nett momentum flux flowing out of a control volume and the rate of change of momentum within the control volume equals to the total of applied forces. We can write this statement as an equation

$$\frac{d}{dt}\int_0^l \rho Q(z,t)\,dz + \alpha\rho Q(l,t)u(l,t) - \alpha\rho Q(0,t)u(0,t) = F, \tag{2.2.1}$$

where $\rho Q(z,t)$ is the momentum, and α is the correction factor of the momentum flux. Then, the total of applied forces is defined as follows

$$F = p(0,t)A(0,t) - p(l,t)A(l,t) + \int_0^l p\frac{\partial A}{\partial z}\,dz + \int_0^l f\,dx, \tag{2.2.2}$$

where f is the blood friction force per unit length. If equation (2.2.2) is substituted to equation (2.2.1), then we obtain

$$\frac{d}{dt}\int_0^l \rho Q(z,t)\,dz + \int_0^l \frac{\partial(\alpha\rho Qu)}{\partial z}\,dz = -\int_0^l \frac{\partial(pA)}{\partial z}\,dz + \int_0^l p\frac{\partial A}{\partial z}\,dz + \int_0^l f\,dz. \tag{2.2.3}$$

Because ρ and l are constant, we can rewrite equation (2.2.3) as follows

$$\int_0^l \left(\rho\frac{\partial Q}{\partial t} + \rho\frac{\partial(\alpha Qu)}{\partial z}\right)dz = \int_0^l \left(-\frac{\partial(pA)}{\partial z} + p\frac{\partial A}{\partial z} + f\right)dz. \tag{2.2.4}$$

Equation (2.2.4) is satisfied for any value of l, so we have

$$\rho\frac{\partial Q}{\partial t} + \rho\frac{\partial(\alpha Qu)}{\partial z} = -\frac{\partial(pA)}{\partial z} + p\frac{\partial A}{\partial z} + f. \tag{2.2.5}$$

Noticing that

$$-\frac{\partial(pA)}{\partial z} = -\mathrm{p}\frac{\partial A}{\partial z} - \mathrm{A}\frac{\partial p}{\partial z}, \tag{2.2.6}$$

we obtain the momentum conservation equation

$$\rho\frac{\partial Q}{\partial t} + \rho\frac{\partial(\alpha Qu)}{\partial z} + \mathrm{A}\frac{\partial p}{\partial z} - f = 0. \tag{2.2.7}$$

2.3. Blood flow models

Based on the mass and momentum conservation laws, we find a one-dimensional blood flow model in human artery

$$\begin{cases} \dfrac{\partial A}{\partial t} + \dfrac{\partial Q}{\partial z} = 0, \\ \dfrac{\partial Q}{\partial t} + \dfrac{\partial}{\partial z}\left(\dfrac{Q^2}{A}\right) + \dfrac{A}{\rho}\dfrac{\partial p}{\partial z} = 0, \end{cases} \tag{2.3.1}$$

where $\alpha = 1$ and $f = 0$. This model (2.3.1) is called the (A, Q) system of blood flow model.

We have $Q = Au$, so the second equation (2.3.1) can be rewritten as follows

$$u\frac{\partial A}{\partial t} + A\frac{\partial u}{\partial t} + 2uA\frac{\partial u}{\partial z} + u^2\frac{\partial A}{\partial z} + \frac{A}{\rho}\frac{\partial p}{\partial z} = 0. \tag{2.3.2}$$

Thus, we get another one-dimensional blood flow model in human artery

$$\begin{cases} \dfrac{\partial A}{\partial z} + \dfrac{\partial(Au)}{\partial z} = 0 \\ \dfrac{\partial u}{\partial t} + \dfrac{\partial}{\partial z}\left(\dfrac{u^2}{2}\right) + \dfrac{1}{\rho}\dfrac{\partial p}{\partial z} = 0. \end{cases} \tag{2.3.3}$$

This model (2.3.3) is called the (A, u) system of blood flow model.

2.4. Finite volume method

Finding the exact solution of a model is not easy, so we shall solve the considered models numerically using a finite volume method with the formulation of the Lax-Friedrichs flux [4].

Notice that the (A, Q) system has two equations and three dependent variables (that is, A, Q, and p), so we define a relation for the blood pressure and the artery cross section area to get two equations with two dependent variables [2, 6]

$$p = p_{\mathrm{ext}} + \beta\left(\sqrt{A} - \sqrt{A_0}\right), \tag{2.4.1}$$

IOP Conf. Series: Materials Science and Engineering **325** (2018) 012014 doi:10.1088/1757-899X/325/1/012014

where A_0 is the artery cross section area when $t = 0$, and p_{ext} is the external pressure. Here, p_{ext} equals to zero, and A_0 is constant. Then, β is a parameter which relates to the wall elastic properties, and it is defined as follows [2]

$$\beta(z) = \frac{4\sqrt{\pi}h_0 E(z)}{3A_0}, \tag{2.4.2}$$

where $E(z)$ is the Young's modulus of elasticity.

For numerical solutions, let us consider the space domain discretisation (as shown in Figure 2) where $\Delta z = z_{i+1/2} - z_{i-1/2}$ or $\Delta z = z_i - z_{i-1}$, and the time domain discretisation $t^n = n\Delta t$ for non-negative integers i and n.

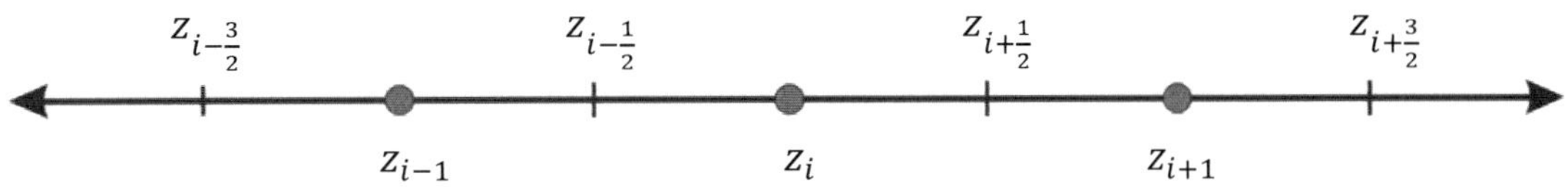

Figure 2. Spatial domain discretisation.

Using algebraic operations [2], we know that

$$\frac{A}{\rho}\frac{\partial p}{\partial z} = \frac{\partial}{\partial z}\left(\frac{\beta A^{\frac{3}{2}}}{3\rho}\right) + \frac{A}{\rho}\frac{d\beta}{dz}\left(\frac{2}{3}\sqrt{A} - \sqrt{A_0}\right), \tag{2.4.3}$$

so the (A, Q) system can be written as a balance law

$$\bar{v}_t + \bar{f}\,(\bar{v})_z = \bar{s}(\bar{v}), \tag{2.4.4}$$

where the quantity, the flux, and the source term are respectively given by

$$\bar{v} = \begin{bmatrix} A \\ Q \end{bmatrix},\ \bar{f}(\bar{v}) = \begin{bmatrix} Q \\ \frac{Q^2}{A} + \frac{\beta}{3\rho}A^{\frac{3}{2}} \end{bmatrix},\ \text{and}\ \bar{s}(\bar{v}) = \begin{bmatrix} 0 \\ \frac{A}{\rho}\frac{d\beta}{dz}\left(\sqrt{A_0} - \frac{2}{3}\sqrt{A}\right) \end{bmatrix}. \tag{2.4.5}$$

Using the finite volume framework, we assume that $\bar{V}_i^n$, $\bar{f}(\bar{V}_i^n)$, and $\bar{S}_i^n$ are approximate values of $\bar{v}(z_i, t^n)$, $\bar{f}(\bar{v}(z_i, t^n))$, and $\bar{s}(\bar{v}(z_i, t^n))$, respectively. Then, we have the following vectors

$$\bar{V}_i^n = \begin{bmatrix} A_i^n \\ Q_i^n \end{bmatrix},\ \bar{f}(\bar{V}_i^n) = \begin{bmatrix} Q_i^n \\ \frac{(Q_i^n)^2}{A_i^n} + \frac{\beta}{3\rho}(A_i^n)^{\frac{3}{2}} \end{bmatrix},\ \text{and}\ \bar{S}_i^n = \begin{bmatrix} 0 \\ \frac{A_i^n}{\rho}\frac{d\beta}{dz}\left(\sqrt{A_0} - \frac{2}{3}\sqrt{A_i^n}\right) \end{bmatrix}. \tag{2.4.6}$$

Thus, the fully-discrete finite volume method [4] for the numerical solution to balance law (2.4.4) is

$$\bar{V}_i^{n+1} = \bar{V}_i^n - \frac{\Delta t}{\Delta z}\left(\bar{F}_{i+\frac{1}{2}}^n - \bar{F}_{i-\frac{1}{2}}^n\right) + \Delta t \bar{S}_i^n \tag{2.4.7}$$

with the formulation of the Lax-Friedrichs flux

$$\bar{F}_{i+\frac{1}{2}}^n = \frac{\bar{f}(\bar{V}_{i+1}^n) + \bar{f}(\bar{V}_i^n)}{2} - \frac{\Delta z}{2\Delta t}(\bar{V}_{i+1}^n - \bar{V}_i^n) \tag{2.4.8}$$

and

$$\bar{F}_{i-\frac{1}{2}}^n = \frac{\bar{f}(\bar{V}_i^n) + \bar{f}(\bar{V}_{i-1}^n)}{2} - \frac{\Delta z}{2\Delta t}(\bar{V}_i^n - \bar{V}_{i-1}^n). \tag{2.4.9}$$

From equations (2.4.7)-(2.4.9), we get the numerical scheme for (A, Q) system (2.3.1) as follows

$$A_i^{n+1} = A_i^n - \frac{\Delta t}{\Delta z}\left(F_{i+\frac{1}{2}}^n - F_{i-\frac{1}{2}}^n\right) \tag{2.4.10}$$

with the formulation of the Lax-Friedrichs flux

$$F_{i+\frac{1}{2}}^n = \frac{1}{2}(Q_{i+1}^n + Q_i^n) - \frac{\Delta z}{2\Delta t}(A_{i+1}^n - A_i^n), \tag{2.4.11}$$

$$F_{i-\frac{1}{2}}^n = \frac{1}{2}(Q_i^n + Q_{i-1}^n) - \frac{\Delta z}{2\Delta t}(A_i^n - A_{i-1}^n), \tag{2.4.12}$$

and

$$Q_i^{n+1} = Q_i^n - \frac{\Delta t}{\Delta z}\left(\mathcal{F}_{i+\frac{1}{2}}^n - \mathcal{F}_{i-\frac{1}{2}}^n\right) + \Delta t\left(\frac{A_i^n}{\rho}\frac{d\beta}{dz}\left(\sqrt{A_0} - \frac{2}{3}\sqrt{A_i^n}\right)\right) \quad (2.4.13)$$

with the formulation of the Lax-Friedrichs flux

$$\mathcal{F}_{i+\frac{1}{2}}^n = \frac{1}{2}\left[\frac{(Q_{i+1}^n)^2}{A_{i+1}^n} + \frac{\beta}{3\rho}(A_{i+1}^n)^{\frac{3}{2}} + \frac{(Q_i^n)^2}{A_i^n} + \frac{\beta}{3\rho}(A_i^n)^{\frac{3}{2}}\right] - \frac{\Delta z}{2\Delta t}(Q_{i+1}^n - Q_i^n), \quad (2.4.14)$$

$$\mathcal{F}_{i-\frac{1}{2}}^n = \frac{1}{2}\left[\frac{(Q_i^n)^2}{A_i^n} + \frac{\beta}{3\rho}(A_i^n)^{\frac{3}{2}} + \frac{(Q_{i-1}^n)^2}{A_{i-1}^n} + \frac{\beta}{3\rho}(A_{i-1}^n)^{\frac{3}{2}}\right] - \frac{\Delta z}{2\Delta t}(Q_i^n - Q_{i-1}^n). \quad (2.4.15)$$

Then, let us consider the (A, u) system in equation (2.3.3) as a conservation law

$$\bar{v}_t + \bar{f}(\bar{v})_z = \bar{0} \quad (2.4.16)$$

where the quantity and the flux are

$$\bar{v} = \begin{bmatrix} A \\ u \end{bmatrix} \text{ and } \bar{f}(\bar{v}) = \begin{bmatrix} Au \\ \frac{u^2}{2} + \frac{p}{\rho} \end{bmatrix}. \quad (2.4.17)$$

Using the finite volume framework, we assume that $\bar{V}_i^n$ and $\bar{f}(\bar{V}_i^n)$ are approximate values of $\bar{v}(z_i, t^n)$ and $\bar{f}(\bar{v}(z_i, t^n))$, respectively. Then, with the same spatial and time discretisation, we have the following vectors

$$\bar{V}_i^n = \begin{bmatrix} A_i^n \\ u_i^n \end{bmatrix} \text{ and } \bar{f}(\bar{V}_i^n) = \begin{bmatrix} A_i^n u_i^n \\ \frac{(u_i^n)^2}{2} + \frac{p_i^n}{\rho} \end{bmatrix}. \quad (2.4.18)$$

Thus, the fully-discrete finite volume method [4] for the numerical solution to conservation law (2.4.16) is

$$\bar{V}_i^{n+1} = \bar{V}_i^n - \frac{\Delta t}{\Delta z}\left(\bar{\mathrm{F}}_{i+\frac{1}{2}}^n - \bar{\mathrm{F}}_{i-\frac{1}{2}}^n\right) \quad (2.4.19)$$

with the formulation of the Lax-Friedrichs flux

$$\bar{\mathrm{F}}_{i+\frac{1}{2}}^n = \frac{\bar{f}(\bar{V}_{i+1}^n) + \bar{f}(\bar{V}_i^n)}{2} - \frac{\Delta z}{2\Delta t}(\bar{V}_{i+1}^n - \bar{V}_i^n) \quad (2.4.20)$$

and

$$\bar{\mathrm{F}}_{i-\frac{1}{2}}^n = \frac{\bar{f}(\bar{V}_i^n) + \bar{f}(\bar{V}_{i-1}^n)}{2} - \frac{\Delta z}{2\Delta t}(\bar{V}_i^n - \bar{V}_{i-1}^n). \quad (2.4.21)$$

From equations (2.4.19)-(2.4.21), we get the numerical scheme for (A, u) system (2.3.3) as follows

$$A_i^{n+1} = A_i^n - \frac{\Delta t}{\Delta z}\left(F_{i+\frac{1}{2}}^n - F_{i-\frac{1}{2}}^n\right) \quad (2.4.22)$$

with the formulation of the Lax-Friedrichs flux

$$F_{i+\frac{1}{2}}^n = \frac{1}{2}(A_{i+1}^n u_{i+1}^n + A_i^n u_i^n) - \frac{\Delta z}{2\Delta t}(A_{i+1}^n - A_i^n), \quad (2.4.23)$$

$$F_{i-\frac{1}{2}}^n = \frac{1}{2}(A_i^n u_i^n + A_{i-1}^n u_{i-1}^n) - \frac{\Delta z}{2\Delta t}(A_i^n - A_{i-1}^n), \quad (2.4.24)$$

and

$$u_i^{n+1} = u_i^n - \frac{\Delta t}{\Delta z}\left(\mathcal{F}_{i+\frac{1}{2}}^n - \mathcal{F}_{i-\frac{1}{2}}^n\right) \quad (2.4.25)$$

with the formulation of the Lax-Friedrichs flux

$$\mathcal{F}_{i+\frac{1}{2}}^n = \frac{1}{2}\left[\frac{(u_{i+1}^n)^2}{2} + \frac{p_{i+1}^n}{\rho} + \frac{(u_i^n)^2}{2} + \frac{p_i^n}{\rho}\right] - \frac{\Delta z}{2\Delta t}(u_{i+1}^n - u_i^n), \quad (2.4.26)$$

$$\mathcal{F}_{i-\frac{1}{2}}^n = \frac{1}{2}\left[\frac{(u_i^n)^2}{2} + \frac{p_i^n}{\rho} + \frac{(u_{i-1}^n)^2}{2} + \frac{p_{i-1}^n}{\rho}\right] - \frac{\Delta z}{2\Delta t}(u_i^n - u_{i-1}^n). \quad (2.4.27)$$

3. Results and discussion

In this section, we simulate the numerical scheme of each system using the MATLAB programming language. For all simulations, we use the same coefficient values, initial and boundary conditions. Then, we investigate the numerical results, and determine which system is better numerically.

3.1. Numerical settings

For the first numerical simulation, we take the length of human artery $l_1 = 15$ cm and $t \in [0,\ 0.035]$ seconds. We assume the Young's modulus E is constant, so it implies that β is also constant. In other words, the values of β and E do not change at every $z \in [0,\ 15]$ and the value of $\frac{d\beta}{dz}$ equals to zero. Moreover, we take $\Delta z = 0.005$ and $\Delta t = 0.002\Delta z$. Methods are stable as long as Δz and Δt satisfy the CFL (Courant-Friedrichs-Lewy) condition [4]. All simulations have the length unit in cm and the time unit in second.

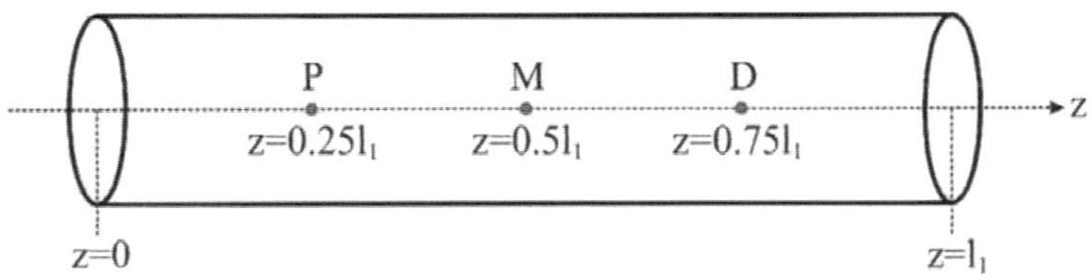

Figure 3. Illustration of human artery at $t = 0$.

Then, the layout of numerical simulation is explained as follows. Figure 3 shows the illustration of human artery at $t = 0$ and $z \in [0,\ 15]$. In this simulation, we set the P, M, and D points as the monitoring points to investigate the pressure variety. These P, M, and D points are proximal, medium, and distal points, respectively. Consider that P point is the nearest point from the heart, and D point is the farthest point from the heart. Table 1 shows the coefficient values used for this numerical simulation (see [2]).

Table 1. The coefficient values for numerical simulation [2].

Coefficient	Value
Blood density, ρ	1 gram/cm^3
Young's modulus, E_0	3x10^6 dyne/cm^2
Artery wall thickness, h	0.05 cm
Initial artery cross section area, A_0	$\pi 0.5^2$ cm^2

After that, we set the initial and boundary conditions to each of A, Q, u, and p. We are given initial values $A(z,0) = A_0$, $Q(z,0) = 0$, $u(z,0) = 0$, and $p(z,0) = 0$ for every $z \in (0,\ 15)$. The boundary values for this simulation are taken as follows. At the left boundary, we give an impulse input in the form of a single sine wave with a small time period,

$$p(0,t) = 10^3 \sin\left(\frac{\pi t}{0.0025}\right). \tag{3.1.1}$$

At the left boundary values of A and Q, let us consider the characteristic variables of (A, Q) system, W_1 and W_2, that have been explained in [2]

$$W_2 = \frac{Q}{A} - 2\sqrt{\frac{2\beta}{\rho}}A^{\frac{1}{4}}, \tag{3.1.2}$$

$$W_1 = W_2 + 4\sqrt{\frac{2}{\rho}}\left(\sqrt{p + \beta\sqrt{A_0}}\right), \tag{3.1.3}$$

$$A = \left(\frac{\rho}{\beta}\right)^2 \frac{(W_1 - W_2)^4}{4^5}, \tag{3.1.4}$$

$$Q = A\frac{W_1 + W_2}{2}. \tag{3.1.5}$$

The left boundary values of A and u are given as follows

$$A = \left(\frac{p + \beta\sqrt{A_0}}{\beta}\right)^2, \tag{3.1.6}$$

$$u = \sqrt{\frac{8\beta}{\rho}}\left(A^{\frac{1}{4}} - A_0^{\frac{1}{4}}\right). \tag{3.1.7}$$

The left boundary value of u in equation (3.1.7) is an equation of the wave characteristic speed (see [6]). Finally, each of right boundary values of A, Q, u, and p equals to the corresponding value of the nearest neighbour in the domain.

3.2. Numerical results

In this section, we show our numerical results. We take $T = 0.035$ seconds.

Numerical results of the (A, Q) system are shown in Figures 4-7. Based on the results that we obtain, with a constant Young's modulus, the artery cross section area is directly proportional to the blood pressure (see Figure 4 and Figure 5). Furthermore, the amplitude of blood pressure decreases with respect to space and time (see Figure 6 and Figure 7). We conclude that it is caused by the dissipation of the numerical method. As long as the solution is continuous, we also conclude that smaller cell width leads to smaller amplitude's decrement.

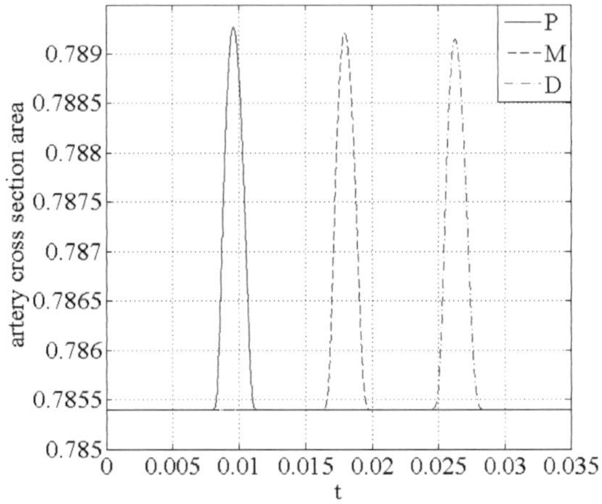

Figure 4. Graphic of artery cross section area of the (A, Q) system with respect to time.

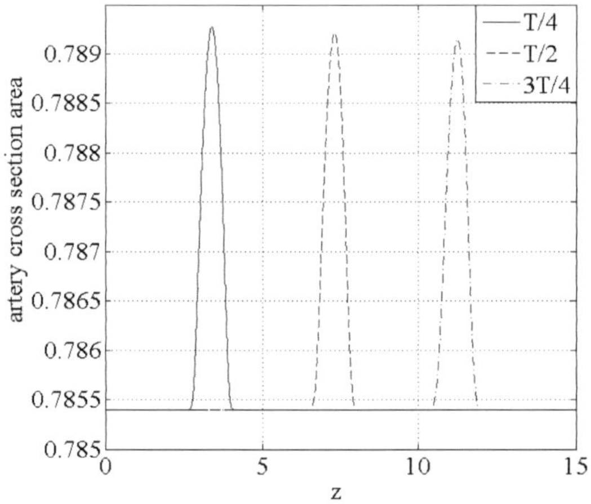

Figure 5. Graphic of artery cross section area of the (A, Q) system with respect to space.

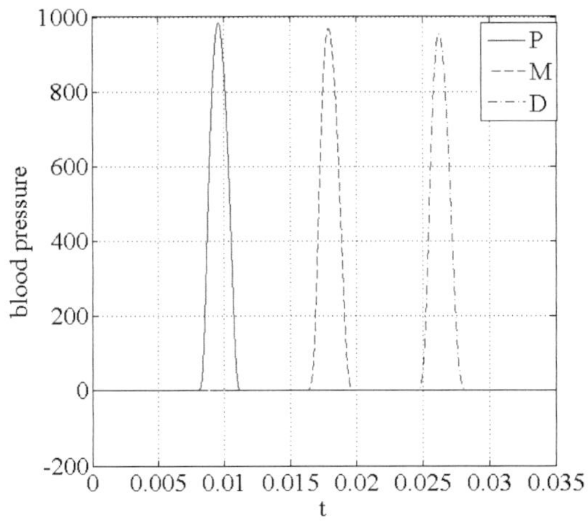

Figure 6. Graphic of blood pressure of the (A, Q) system with respect to time.

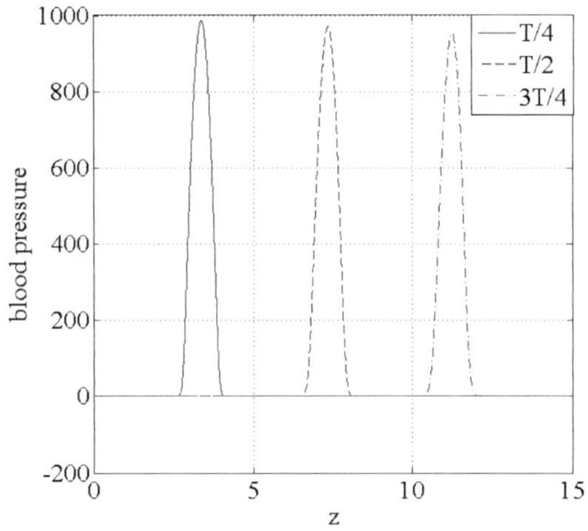

Figure 7. Graphic of blood pressure of the (A, Q) system with respect to space.

Numerical results of the (A, u) system are illustrated in Figures 8-11. Like the previous discussion, the artery cross section area is directly proportional to the blood pressure (see Figure 8 and Figure 9). Furthermore, the amplitude of the blood pressure decreases with respect to space and time (see Figure 10 and Figure 11). Again we conclude that it is caused by the dissipation of the numerical method. As long as the solution is continuous, we also conclude that smaller cell width results in smaller amplitude's decrement.

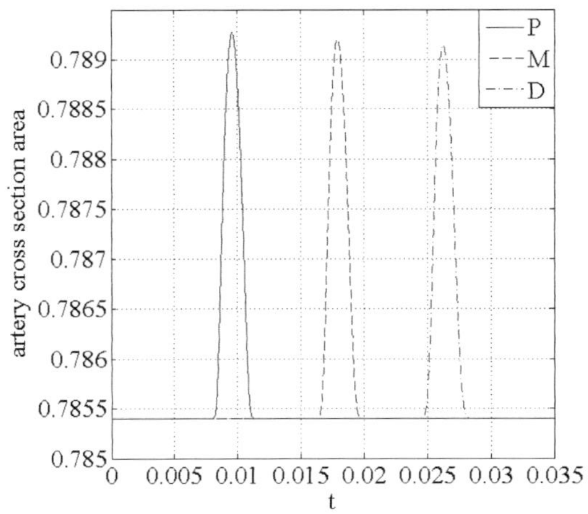

Figure 8. Graphic of artery cross section area of the (A, u) system with respect to time.

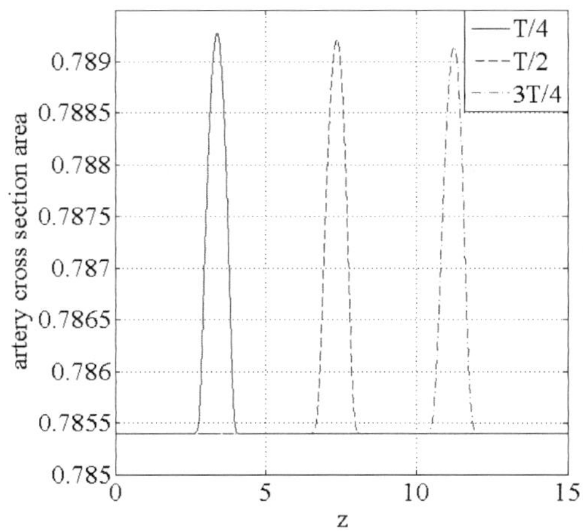

Figure 9. Graphic of artery cross section area of the (A, u) system with respect to space.

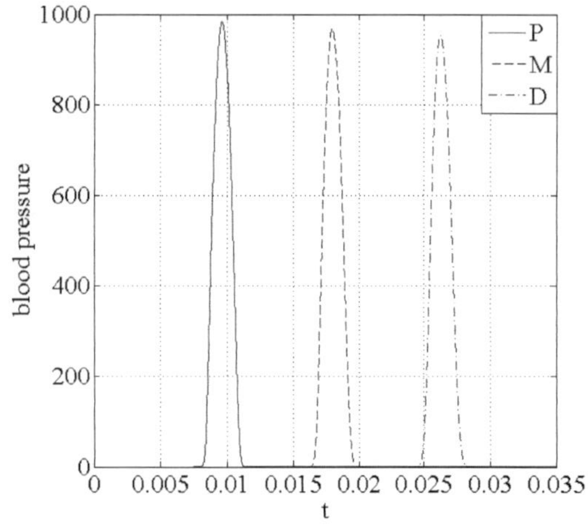

Figure 10. Graphic of blood pressure of the (A, u) system with respect to time.

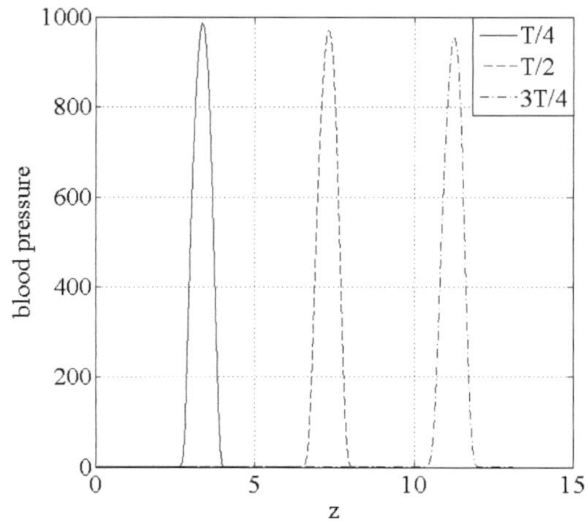

Figure 11. Graphic of blood pressure of the (A, u) system with respect to space.

The numerical results of the (A, Q) system and those of the (A, u) system look similar. However, there is a difference between those numerical results. The blood pressure of the (A, Q) system has negative values (see Figure 12), whereas the values of the blood pressure of the (A, u) system are non-negative (see Figure 13).

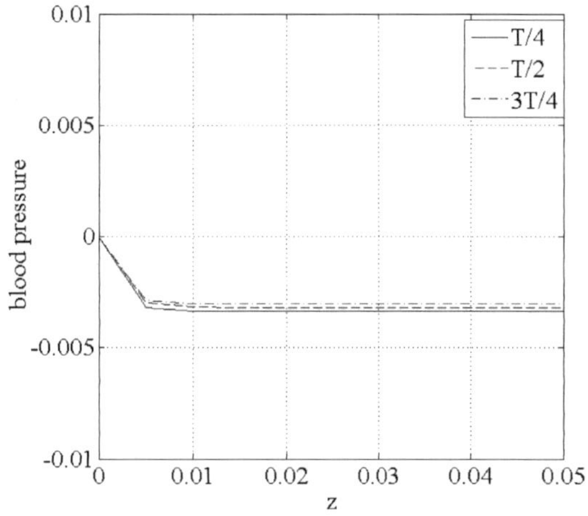

Figure 12. Magnification of blood pressure's graphic of the (A, Q) system with the space.

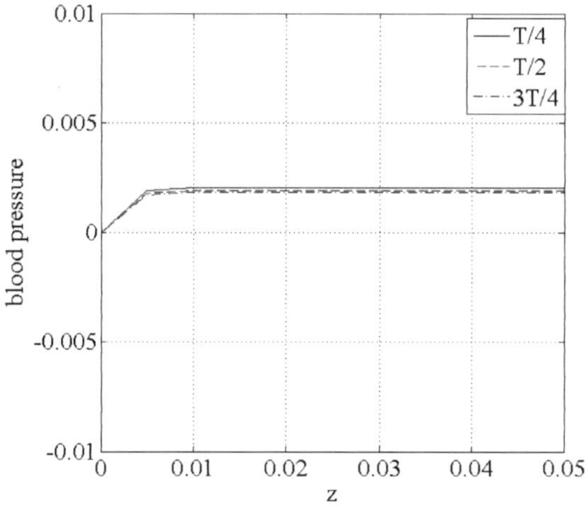

Figure 13. Magnification of blood pressure's graphic of the (A, u) system with the space.

3.3. Residuals

In this subsection, we compute the residual of A for each system to determine which system is better numerically. A system is said to be better numerically, if its residual is closer to zero.

We compute the residual of A with this formula [5]

$$\begin{aligned} E_{i+1/2}^{n-1/2} &= \frac{\Delta z}{2}\left[q_i^n - q_i^{n-1} + q_{i+1}^n - q_{i+1}^{n-1}\right] \\ &+ \frac{\Delta t}{2}\left[f\left(q_{i+1}^{n-1}\right) - f\left(q_i^{n-1}\right) + f\left(q_{i+1}^n\right) - f\left(q_i^n\right)\right], \end{aligned} \tag{3.3.1}$$

where in this formula (3.3.1) we take $q_i^n = A_i^n$ to each of system, $f(q_i^n) = Q_i^n$ to the (A, Q) system, and $f(q_i^n) = A_i^n u_i^n$ to the (A, u) system.

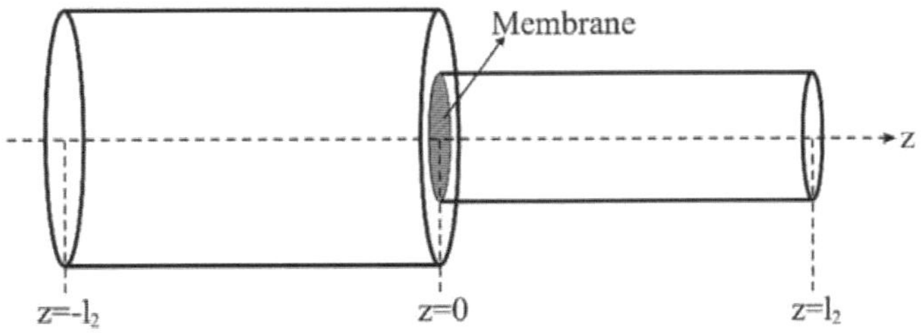

Figure 14. Illustration of a human artery with a membrane at $t = 0$.

The numerical layout to compute the residual of both systems is as follows. We assume to have human artery in the interval $[-l_2, l_2]$ with $l_2 = 2$. The initial condition is shown in Figure 14. There is a membrane at the artery cross section $S(0,0)$ at time $t = 0$. Then, at time $t > 0$, the membrane is removed completely, so blood can flow through.

Let us consider initial and boundary conditions for this simulation as follows. We are given a discontinuous initial function A

$$A(z,0) = \begin{cases} 10, \text{ if } z \leq 0, \\ 5, \text{ if } z > 0, \end{cases} \tag{3.3.2}$$

for all $z \in (-2, 2)$. The initial values of Q, u, and p are zero, for all $z \in (-2, 2)$. Then, we are given the left and right boundary values of A

$$A(-2, t) = 10, \tag{3.3.3}$$

$$A(2, t) = 5, \tag{3.3.4}$$

for all $t > 0$. The left and right boundary values of Q, u, and p are zero for all $z \in (-2, 2)$. The time domain and coefficient values are taken the same as the previous simulation.

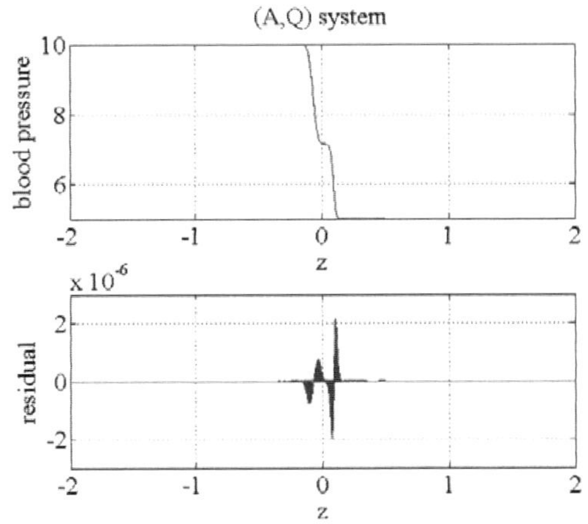

Figure 15. Blood pressure and residual of A of the (A, Q) system.

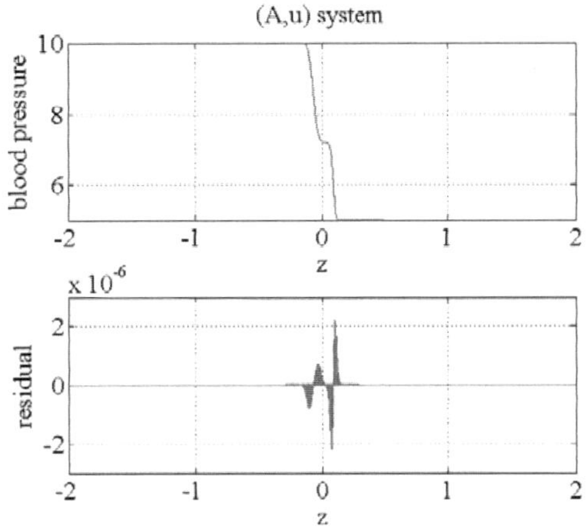

Figure 16. Blood pressure and residual of A of the (A, u) system.

Residuals of A are plotted in Figure 15 and Figure 16, for (A, Q) and (A, u) systems, respectively. The blood pressure's graphic looks alike, because the difference of the residual is small. Figure 15 shows that the lower bound of the residual of A using the (A, Q) system is larger than or equal to -2×10^{-6}. Figure 16 shows that the lower bound of the residual of A using the (A, u) system is smaller than

-2×10^{-6}. The upper bound of the residual of A using both systems look similar, that is larger than 2×10^{-6}.

4. Conclusion

Using our finite volume numerical method, the blood pressure pulse propagates from high pressure to lower one, as expected. The form of the propagating pulse does not change, but the amplitude decreases due to the dissipation of the numerical method. We conclude that the (A, Q) system is better numerically than the (A, u) system, because the absolute of the residual of A using the (A, Q) system is smaller than the absolute of the residual of A using the (A, u) system. Our results are limited to one-dimensional problems, but could be extended for multi-dimensional problems in future research.

References

[1] Budiawan I W and Mungkasi S 2017 Finite volume numerical solution to a blood flow problem in human artery *J. of Phys.: Conf. Series* **795** p 012042.

[2] Formaggia L, Nobile F and Quarteroni A 2002 A one dimensional model for blood flow: application to vascular prosthesis *Mathematical Modeling and Numerical Simulation in Continuum Mechanics* **19 p** 137.

[3] LeVeque R J 1992 *Numerical Methods for Conservation Laws* (Basel: Springer).

[4] LeVeque R J 2002 *Finite Volume Methods for Hyperbolic Problems* (Cambridge: Cambridge University Press).

[5] Mungkasi S, Li Z and Roberts S G 2014 Weak local residuals as smoothness indicators for the shallow water equations *Applied Mathematics Letters* **30** p 51.

[6] Sherwin S J, Franke V, Peiro J and Parker K 2003 One-dimensional modelling of a vascular network in space-time variables *J. of Eng. Math.* **47** p 217.

Web Application to Monitor Logistics Distribution of Disaster Relief Using the CodeIgniter Framework

Mohamad Jamil[1], Mohamad Ridwan Lessy[2*])

[1)] Departement Informatics Engineering, Faculty of Engineering, Universitas Khairun .Jl. Pertamina Kampus II Unkhair Gambesi Kota Ternate, Maluku Utara, Indonesia 97719

[2)] Departement Marine Science Study Program, Faculty of Fisheries, Universitas Khairun Jl. Pertamina Kampus II Unkhair Gambesi Kota Ternate, Maluku Utara, Indonesia 97719

Abstract. Disaster management is the responsibility of the central government and local governments. The principles of disaster management, among others, are quick and precise, priorities, coordination and cohesion, efficient and effective manner. Help that is needed by most societies are logistical assistance, such as the assistance covers people's everyday needs, such as food, instant noodles, fast food, blankets, mattresses etc. Logistical assistance is needed for disaster management, especially in times of disasters. The support of logistical assistance must be timely, to the right location, target, quality, quantity, and needs. The purpose of this study is to make a web application to monitorlogistics distribution of disaster relefusing CodeIgniter framework. Through this application, the mechanisms of aid delivery will be easily controlled from and heading to the disaster site.

Keywords - Application; Disaster Monitoring; Web; framework; Codeigniter.

1. Introduction

Disaster is an event or series of events that threaten and disrupt the life and livelihood of the community caused either by natural factors and / or non-natural factors as well as human factors resulting in the occurrence of human casualties, environmental damage, property loss, and psychological effects [1]. Disaster management is the responsibility of central and local governments. The principles in disaster management are fast and precise, priority, coordination and integration, efficient and effective. In relation to that matter, to reduce the risk of the disasterimpact, disaster management is needed.

Disaster Management is required to prevent and reduce the losses arising from disasters that occurred, either in the form of property or material losses, and ensuring adequate assistance to victims of natural disasters before, during and after the disaster [2]. In a state of emergency disaster, people are in dire need of help from government, community or private organizations. The assistancesthat most people need are logistical support, the assistance that covers the daily needs of the community, such as food, instant noodles, fast food, blankets, mattresses and others. Logistic assistance is needed for disaster management especially when the disaster occured. The support of logistical support must be timely, to the right location, target, quality, quantity, and need.

Nurhayati and Lucky Feliciono Waha in their research argue that the modeling of Application modeling can be used to monitorthe distribution of aid, by collecting, processing and presenting the location, victims, and the needs of the victims, and can see whether the assistance has been received or not. [3]

Indah Permata Sari in her research argues that in order to smooth and acceleratethe collecting information data about disaster victim handling,a special information system that can accommodate all data of disaster victim handling is required[4]

Lucky Feliciano Waha in his research argued that the lack of disaster-related information that occurred cause the process of distributing aid to the community is not on target[5]

Nicky Nia Gustriani in her research argued that the success of government in tackling the disaster that occurs depends on how the implementation of disaster management system information. Because to make the right decision, the government needs fast and accurate data shortly after the disasteroccurred.[6]

The hierarchical procedure demands a process that should be able to run quickly but become a bit slow because the upnormal condition. Therefore, the procedures that can be processed quickly, accurately, can be accounted for and ensure that public services relating to the management of mechanisms of aid distributioncan be channeled well and well targeted is required. One the ways is to develop a monitoring application of logistics assistance distribution web-based with the concept of framework .

A simple framework can be defined as a collection of functions / procedures and classes for a particular purpose that is ready to be used so as to simplify and speed up a programmer's work, without having to create a function or class from the beginning [7]

One of the concepts adopted by the framework is MVC (Model, View, Controler). MVC separates application development based on major components that develop an application such as data manipulation, user interface, and parts that become application controls.

Based on the description of existing problems, then in this research the researcher will make a Monitoring Application of Disaster Logistics Aid Distribution using codeignetir framework. It is expected that this research will be able to assist the community and also related offices in distributing and monitoring logistic support from and to disaster location more quickly and on target.

2. Research Method

One of the method used in this research is waterfal model. It is called linear sequential model or software life sequence sequentially or sorted starting from analysis, design, coding, testing, and maintenance [8]. There are 5 stages: needs analysis, system design, programming stage, implementation and evaluation.

The first stage is the analysis of system requirement phase. The requirements are divided into 2 ie functional and non-functional requirements. Functional requirements include user-level sharing and the classification of system capability such as logistics requests and logistics approvals. Non-functional requirements include devices that support the application. The second stage is the system design. System design includes interface design and the design of program workflow. The focus at this stage is the creation of interface design that facilitates users in using the system.

The third stage in this research is the programming stage. At this stage, the process of translation of the system workflow into the PHP programming language into the codeignetier framework is conducted.

The fourth stage in the development of this system is the implementation stage. Here, the implementation stage is at the level of testing according to the needs of existing systems in the monitoring application of distribution of logistics aid for disaster. From this stage, it can be known whether the results shown by the tool is in accordance with the initial plan so that in this stage the researcher can simultaneously run the evaluation phase.

3. Finding and discussion

The system design aims to find the optimal form of the system that will be developed by considering the problems and needs. This application is designed on PHP programming by using codeigniter framework. In testing this application, it runs on localhost with local server XAMPP 1.6.7

3.1. System Analysis

Disaster management in North Maluku Province can not be carried out by the North Maluku provincial government alone, but other actors are also needed to assist the sustainability of disaster management. The

IOP Conf. Series: Materials Science and Engineering **325** (2018) 012015 doi:10.1088/1757-899X/325/1/012015

actors consist of Sub-district Apparatus, Village Apparatus, and Community. Cooperation and coordination between these stakeholders can facilitate the implementation of aid distribution mechanism. Village officials (lurah) coordinate with sub-district and village apparatuses to carry out the mechanism of distribution and take good measures to handle the flood disaster, coordination among stakeholders is considered to fulfill what should be done by the Public Administration Agency. The process of aid distribution based on SOP can be seen in Figure 1 below:

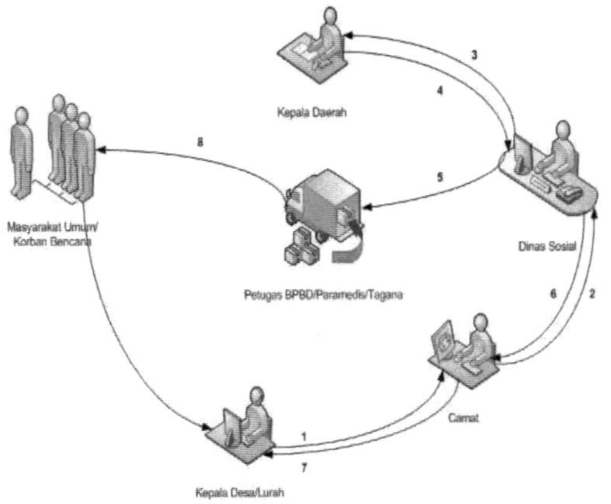

Figure 1 Illustration Standard operational Procedure (SOP) running System

3.2. Data Flow Diagrams

3.2.1. Data Flow Diagram Level 0

DFD Level 0, describes the system globally, contains only one system process that is connected to two entities whichaims to provide an overview to the analyst systems and programmaker about the input into a process that finally produce an output. As for DFD level 0 designs the monitoring application of logistics distribution of disaster relief can be seen in figure 2

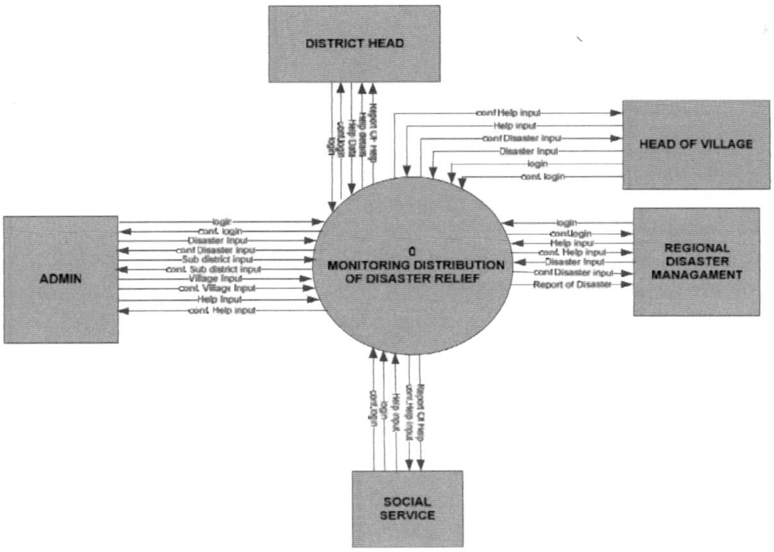

Figure 2 DFD Level 0 Monitoring Application of Logistics Distribution Of Disaster Relief

IOP Conf. Series: Materials Science and Engineering **325** (2018) 012015 doi:10.1088/1757-899X/325/1/012015

Figure 2 defines each user who can later use the monitoring application of natural disaster relief distribution, ie Social Service, Regional Disaster Managament, District Head, Head Of Village, and Administrator.

3.3. Flowchart Design

The flow of flowchart system design is very important in making a computer program. There are several ways to implement a program, one of them is by using flowchart.

Figure 3 Flowchart System

3.4. ***Database Design***

Database design is a series of specific statements that relate to various data processing, such as data objects that will be processed by the system, the composition of each data object and attributes that describe and how the relationship between each data object [9]

1. Entity relationship Diagram
2. ERD (Entity Relationship Diagram) is a notation used to perform data modeling activities. The attributes of each data object are written on the Entity relationship Diagram. It can be described by using the decryption of data objects

3.5. ***Menu Structure Design***

The menu structure for the admin is described in Figure 5. The admin manages the system user and its permissions with multiple menus. The menu consists of Webadmin (manage the user), MenuManage, News, Village / Subdistrict, District / City, Help, Logistics and Logout

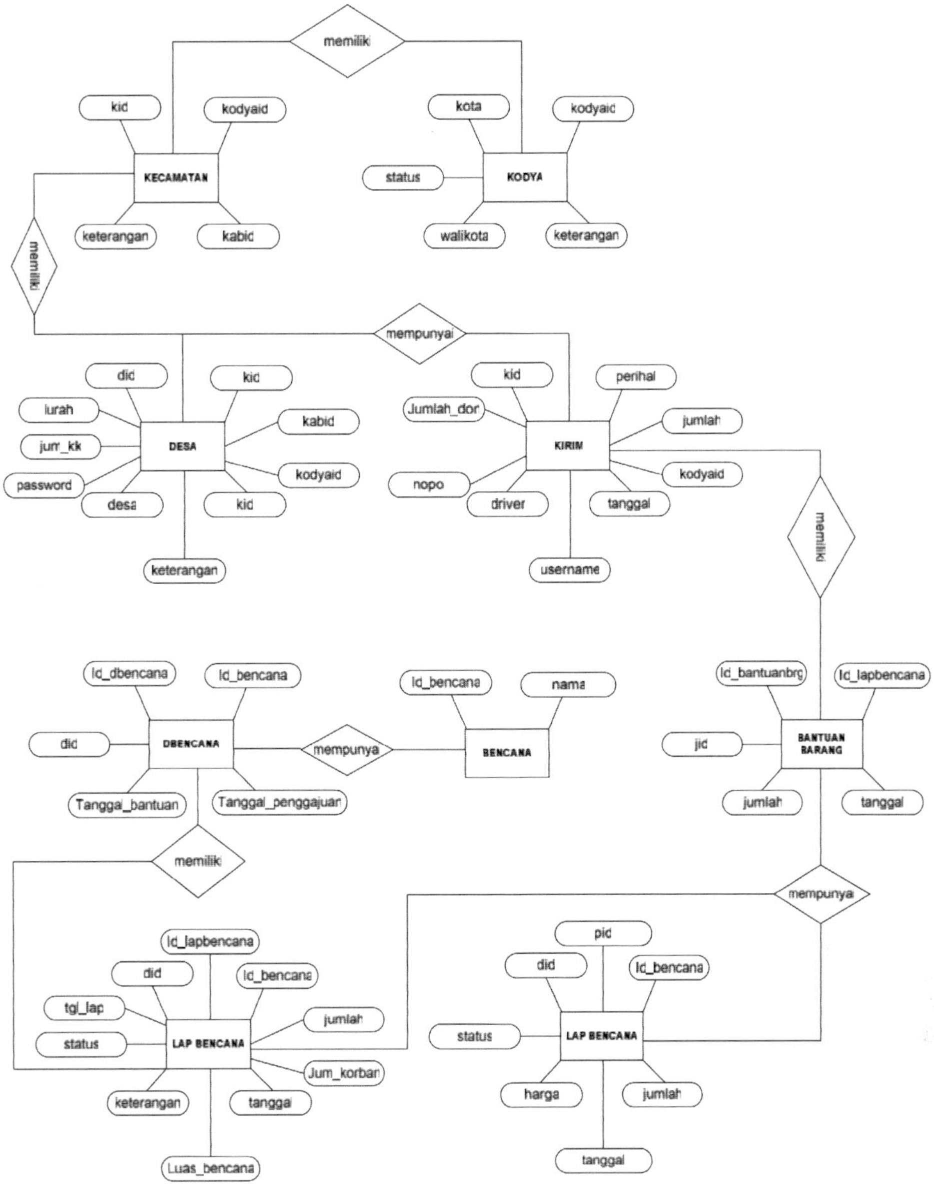

Figure 4 Entity Relationship Diagram

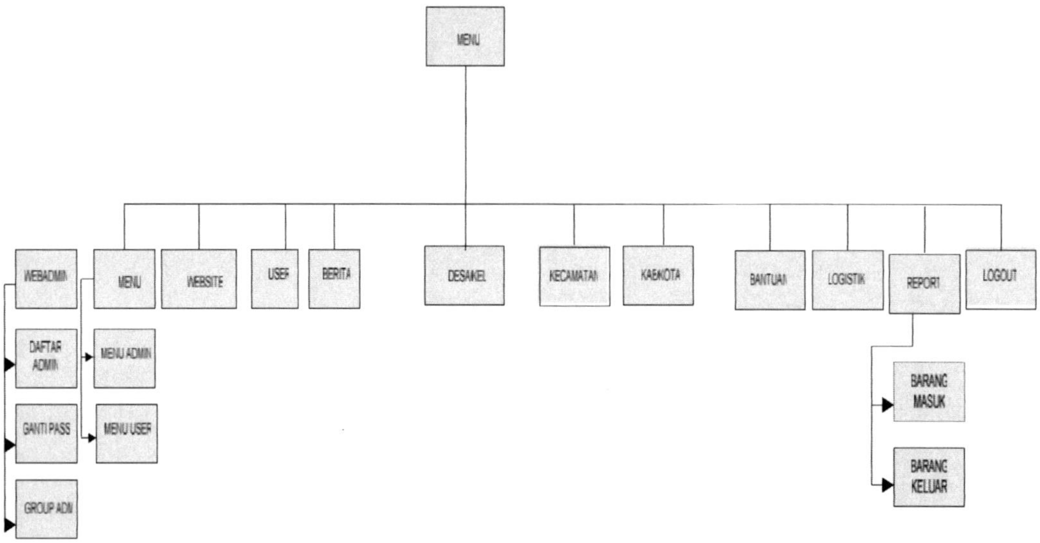

Figure 5 Admin Menu Structure

4. IMPLEMENTATION AND SYSTEM TESTING

4.1. Page for Login User

When a user enters the system, the user is confronted with adisplay of system index of the login page. On the login page, the user is asked to enter a username and password. This page acts as the verification for the incoming user. This login system also has permissions automation based on the user level who login to the page.

Figure 6 User Page Login

The page to input the help approval is a page form used to see the approval process for help

Figure 7 Help Approval Page

5. Conclusion

Based on the tests that are conducted on the application, it can be concluded that the monitoring application of logistic distribution of disaster relief is designed with codeigniter framework that can help the process of disaster data management and the distribution of disaster relief to the destination location can be monitored and managed properly. This system is ready to be implemented in BPBD of North Maluku Province and Social Service of North Maluku Province

Acknowledgments

The researchers would like to thank to the Disaster Management Agency (BPBD) of North Maluku Province who have collaborated on this research. Moreover, the researchers also would like to thank to the Laboratory of Faculty of Engineering in Universitas Khairun that have provided computer network facilities during the study took place.

References

[1] Deputi Bidang Pembinaan Sarana Teknis dan Peningkatan Kapasitas 2007 *Analisis Potensi Rawan bencana alam di Propinsi Maluku dan Papua* (Jakarta: Kementrian Lingkungan Hidup).
[2] Ischa S M, Sarwono, and Imam H 2014 *Tata Kelola Distribusi Bantuan Logistik Korban Bencana Alam (Studi Empiris pada bencana banjir di Kabupaten Bojonegoro)* (Malang: Universitas Brawijaya).
[3] Sri N and Lucky F 2012 Pemodelan Distribusi bantuan bencana berbasis Website *Jurnal Sistem Komputer Unikom* (Bandung).
[4] Permatasari I 2008 *Sistem Informasi Bantuan Penanggulangan Bencana Dinas Sosial Palembang dengan Menggunakan Delphi 2007 dan SQL Server 2008* (Palembang: STIKOM PalComTech).
[5] Waha F and Lucky 2013 *Monitoring Distribusi Bantuan Bencana di Dinas Sosial Provinsi Jawa Barat* (Bandung: UNIKOM).
[6] Gustriani N N 2011 *Penerapan Sistem Informasi Penganggulangan Bencana di BPBD Kota Padang-Skripsi* (Padang: Universitas Andalas).
[7] Hasanuddin 2012 *Pemrograman Web Berbasis Framework.* (Yogyakarta: Universitas Ahmad Dahlan).
[8] Rosa M S 2012 *Modul Pembelajaran Rekayasa Perangkat Lunak (Terstruktur dan Berorientasi Obyek)* vol 53 no 1 (Bandung: Modula).
[9] Fathansyah 2012 *Basis Data* (Bandung: Informatika).
[10] Purnomo H and Sugiantoro R 2010 *Manajemen Bencana: Respons dan Tindakan Terhadap Bencana.* (Yogyakarta: Media Pressindo).

IOP Conf. Series: Materials Science and Engineering **325** (2018) 012016 doi:10.1088/1757-899X/325/1/012016

Data Entities and Information System Matrix for Integrated Agriculture Information System (IAIS)

Halim Budi Santoso[1]; Rosa Delima[2]

[1] Information System Department, Faculty of Information Technology, Duta Wacana Christian University, Yogyakarta

[2] Informatics Department, Faculty of Information Technology, Duta Wacana Christian University, Yogyakarta

Abstract. Integrated Agriculture Information System is a system that is developed to process data, information, and knowledge in Agriculture sector. Integrated Agriculture Information System brings valuable information for farmers: (1) Fertilizer price; (2) Agriculture technique and practise; (3) Pest management; (4) Cultivation; (5) Irrigation; (6) Post harvest processing; (7) Innovation in agriculture processing. Integrated Agriculture Information System contains 9 subsystems. To bring an integrated information to the user and stakeholder, it needs an integrated database approach. Thus, researchers describes data entity and its matrix relate to subsystem in Integrated Agriculture Information System (IAIS). As a result, there are 47 data entities as entities in single and integrated database.

Keywords: Integrated Agriculture Information System, Data Entity, Information System Matrix

1. Introduction

Agriculture is one of the important sectors in Indonesia. Development of Technology in agriculture sector as a government research focus during 2015 – 2045 (Kementrian Riset, Teknologi, dan Pendidikan Tinggi, 2016). Information Technology as one of technologies that is able to support food self-sufficiency through improvement in quality and quantity of agriculture product. This technology is able to support improvement in economic added value for agriculture sector. Implementation of Information System in Agriculture has been implemented in Kenya (Rees, et. al., 2000), Croatia (Renko, Nikolasevic, and Pavicic, 2002), China (Wet, et. al., 2007). Farmers are also has been realized that the usage and importance of implementation Information Technology in Agriculture. 76.81% of farmers agree that Information and Communication Technology (ICT) should be optimized in Agriculture sector and they are ready to be trained the usage of ICT in Agriculture sector. This will be a good motivation to implement ICT in Agriculture Sector. Implementation of ICT for agriculture also has been supported by a good communication infrastructure in rural area, such as telephone network, ISP, and wireless connection.

Researchers are conducting study to build Integrated Agriculture Information System (IAIS). IAIS is a system that is developed to process data, information, and knowledge in Agriculture sector. This system has some subsystems which are connected and linked with integrated database. To build IAIS, researchers developed architecture using Enterprise Architecture Method. TOGAF Enterprise Architecture is chosen to build the blueprint and architecture. One elements of TOGAF Enterprise Architecture is Data Architecture. Data architecture is an approach which is needed to understand and address data management issues. TOGAF data architecture is an element that describe the structure of an organization's logical and physical data assets and management resources. A structured and comprehensive approach to data management enables the effective use of data to capitalize on its competitive advantage. This paper proposes the development of data architecture for integrated agriculture information system. This architecture is based on TOGAF framework which implements holistic approach starting from its organization, its function, its service, and its goal.

2. Literature Review

2.1. e-Agriculture

Development of Information and Communication Technology is perceived not only by urban community but also rural community. In rural area, development of Information and Communication Technology can be related with area economic development. Furthermore, open access to Information and Communication Technology will be opened rural area to market access. Development of Information and Communication Technology in rural area is able to help local business to be more competitive by increasing production capacity. Thus, development of ICT will bring more opportunity by reduce production cost and increasing investment. Information and Communication Technology also bring a positive impact to agriculture sector, especially in rural area. Implementation of ICT in agriculture sector raised and introduced for the first time during World Summit on Information Society in 2003 (Kamran, et. al., 2016). In this session, e-Agriculture is introduced to public. E-Agriculture has a purpose to increase farmer's standard of living (Kamran, et. al., 2016).

E-Agriculture is also able to increase standard of living by reducing cost production in agriculture sector, minimize pre-cultivation process and time, and increase the crops. Thus will bring a financial improvement by increasing the farmer's profit. Farmers are easier to access information that will be useful for agriculture sector. Development of e-Agriculture has some obstacles. Masiello-Riome, et. al. (2008) described that 57% respondent has no attention to e-Agriculture. Farmers are still busy and do not have willingness to change the way they cultivate and process the field. High cost in accessing the Information and Communication Technology is also a challenge. Government should develop less cost information access for rural area.

Information access is an important issue for e-Agriculture development. Farmers need following information: (1) Technique and Practise in agriculture (57%); and (2) Market information (33%) (Masiello-Riome, et. al., 2008). Thus, infrastructure is needed to prepare information access. In Yogyakarta, available infrastructure for ICT is telephone network, ISP, and wireless network (Delima, 2016). Implementation e-Agriculture has been integrated with agriculture operation to ease farmer's job, reduce processing time, minimize cost, and prevent environment from hazardous material. Implementation of e-Agriculture with Mobile Driven Extension has been implemented in India and Kenya (Brugger, 2011) by using Information Technology to collaborate with government, research centre, university, and farmer through internet. This model is supported by applying Call Center approach. By using tele-center, user will be helped by an agent. This agent will provide information that user needed, such as: irrigation, pest management, and other information. Research centre, university, and government should be able to update the information and solve user's problem (Brugger, 2011).

E-Agriculture also provides information about market and the usage of web technology extensively (Kamran, et. al., 2016). Farmer needs following information: (1) Fertilizer price; (2) Agriculture technique and practise; (3) Pest management; (4) Cultivation; (5) Irrigation; (6) Post harvest processing; (7) Innovation in agriculture processing (Delima, 2016). This information is supporting Milavanovic (2014) statement that these information is required by farmers: (1) Information on crops: Farmers need some information related with crops and plants; (2) Information on production techniques; (3) Information on production equipment and agricultural inputs to help farmers to know the tools for soil and plant processing; (4) Market information is an information to support farmers selling the crop during harvest time. These information will be helped to design the data architecture by using TOGAF Framework.

2.2. Data Architecture

The number of data for the enterprise are growing. It encourages the enterprise to depend on data, information, and knowledge in order to run the system and gain the benefit from the system. Thus, it needs to manage data and extract the data in order to get the information and knowledge from the data. Integrated Agriculture Information System as a big system also needs data architecture in order to govern the data. Collection, management, and effective use of data, information, and knowledge are essential ingredients to success of IAIS implementation.

Data architecture comprises data structure and its relationship, principle and guideline governing the design and evolution of data overtime (Hoven, 2004). Data architecture for a business enterprise provides the blueprints for the enterprise's data resources by providing the vision, principles, and standards to create, use, and manage it. (Hoven, 2004). Dunn and Grabski (2001) stated the same representation of the data architecture. Data architecture represent how the data stored logically and physically, how data is organized for information system purpose, and data's correspondence with changing tasks are important to task performance (Dunn and Grabski, 2001). Data architecture also provides framework to enable an enterprise to resolve key issued associated with the quality, timeliness, commonality, and accessibility of the data within the enterprise, and to improve the flow of data between systems within the enterprise (Hoven, 2003). Data architecture also provides a guideline not only for transaction purpose but also to help to decide something related with enterprise strategic solution. Thus, data architecture as a part of enterprise architecture should be defined well as a foundation to build information system. TOGAF as an enterprise architecture standard also has guideline to manage and govern enterprise data. A structured and comprehensive approach to data management enables effective use of data to capitalize on its competitive advantages. Some processes in order to develop data architecture: (1) collect data-related models from existing Business Architecture and Application Architecture materials; (2) rationalize data requirement and align with any enterprise data catalogues and models; (3) Update and develop matrices across the architecture; (4) elaborate data architecture views (http://pubs.opengroup.org/architecture/togaf9-doc/arch/chap10.html)

3. Research Methodology

In conducting this research, researchers followed methodologies:

3.1. Initial Study of ICT Development in Agricultural Sector

Initial study was conducted in 2015 by interviewing 77 farmers from Yogyakarta. From this initial study, it found: (1) 75 % farmers in Yogyakarta are 40 years old and graduated from Senior High School; (2) Farmers use smartphone to look information about agriculture; (3) Farmers lack of IT literacy but they have willingness to use the Information Technology in order to increase their productivity; (4) Farmers need information about prices, farming tools and technique, product marketing and sales, land processing technique, and post-harvest product processing (Delima and Purwadi, 2015).

3.2. Developing Vision Architecture, Business Architecture

After identify the farmers structure and required information, researchers starting to develop vision architecture and business architecture (Delima, Santoso, and Purwadi, 2016; Santoso and Delima, 2016; Delima, Santoso, and Purwadi, 2017). From the architecture vision, it determined that supporting organization for Integrated Agriculture Information System is study group. This organization has the following vision Achieving Farmers Welfare through Precision Farming Based on ICT. It has three missions: (1) providing services to develop farmer capacity using information and communication technology; (2) develop Indonesian IAIS for precision farming; (3) establish online community for every agriculture stakeholders (Delima, Santoso, and Purwadi, 2016). Stakeholders for Integrated Agriculture Information System are categorized into four categories: (1) Farmers; (2) Researchers; (3) Public Sector; and (4) Business sector (Santoso and Delima, 2016). Every stakeholders are analysed and grouped into several group based on its power and interest which reflect the participation of each stakeholders in the development of IAIS and interest in implementation of IAIS (Santoso and Delima, 2016).

Developing vision architecture is followed by developing business architecture. From the business architecture, it is determined that IAIS has four main functions: (1) System, Information Technology Infrastructure, and Organization Standardization; (2) Developing, managing, and maintaining application and infrastructure; (3) IAIS Implementation; (4) Management and Organization. (Delima, Santoso, and Purwadi, 2017). There are 10 available services for IAIS: (1) User Registration; (2) Requesting application services; (3) Requesting ICT Services; (4) Tendering services; (5) Requesting services; (6) Requesting information; (7) Information sharing; (8) Consulting; (9) Recording and observing user specific data; (10) Giving agriculture information and knowledge (Delima, Santoso, and Purwadi, 2017).

3.3. Data requirement for Integrated Agricultural Information System

Delima and Purwadi (2015) studied that farmers need information about prices, farming tools and techniques, product marketing and sales, land processing techniques, and post harvesting processing. Farmers and stakeholder in agriculture sector needs some following information: (1) Fertilizer price; (2) Agriculture technique and practise; (3) Pest management; (4) Cultivation; (5) Irrigation; (6) Post harvest processing; (7) Innovation in agriculture processing. Thus, Integrated Agricultural Information System is able to store data about farmers data, farmers community data, farmers farming activity, daily price, farming technique and processing which will be included in the training session and accommodate by expertise from academician, e-commerce as a place to buy and sell agriculture product.

3.4. Set Up Data Architecture for Integrated Agricultural Information System

In this paper, two components of Integrated Agricultural Information System Data Architecture will be discussed, including: (1) Data entity / Data Component Catalogue; (2) and System / Data Matrix.

4. Analysis and Discussion

4.1. Data Entity / Data Component Catalogue

Data entity is a part of TOGAF data architecture. This data entity identify and maintain a list of all data use across IAIS. Data entity also brings clear definition of where data is stored, either physically or logically.

Table 1. Data Entity / Data Component Catalogue

Data Entity	Description	Logical Data Component	Data Category	Physical Data Component
User	Save and record master user data	Master_user	Master	Disk1
Master Category	Save and record master category data	Master_kategori	Master	Disk1
User Category	Save and record master user category data	Master_user_kat	Master	Disk1
Log User Transaction	Save and record user activity transaction log	Log_user_trans	Log	Disk1
Organization user	Save and record personnel data that are involved in Agriculture Information System Study Group	Master_user_org	Master	Disk1
Organization unit	Save and record organization unit data in Agriculture Information System Study Group	Master_org_unit	Master	Disk1
Forum category	Save and record discussion forum category data	Master_kategori_topik	Master	Disk1
Forum topics	Save and record discussion forum topic	Trans_topik_diskusi	Transaction	Disk2
Forum Comments	Save and record forum comments	Trans_komentar_diskusi	Transaction	Disk2
Farmers Community	Save and record farmers community master data	Master_kel_tani	Master	Disk1
Farmers Community Organization Structure	Save and record personnel data that are involved in management farmers community	Trans_struk_org	Transaction	Disk2
Farmers	Save and record farmers personnel data	Master_petani	Master	Disk1
Farmers membership	Save and record farmer's membership in the community / union	Trans_ang_petani	Transaction	Disk2
Crop Morphology	Save and record crop morphology	Master_morf_tanaman	Master	Disk1
Crop Species	Save and record crop species	Master_spesies_tanaman	Master	Disk1
Farming Activity	Save and record farming activity master data	Master_aktivitas	Master	Disk1
Species Activity	Save and record farming activity and its relation with each species	Master_aktivitas_spesies	Master	Disk1
Farmers Activity	Save and record farming activities that are done by farmers	Trans_petani_aktivityas	Transaction	Disk2
Yields	Save and record farmer's yield / harvest	Trans_hasil_panen	Transaction	Disk2
Cropping Calendar	Save and record cropping calendar data	Trans_kalender_tanam	Transaction	Disk2
Tool, material, and Equipment Category Data	Save and record tool and equipment category data	Master_kategori_alatbahan	Master	Disk1
Farming Tools	Save and record farming tools	Master_alat_tani	Master	Disk1
Farming materials	Save and record farming materials	Master_bahan_tani	Master	Disk1
Supplier Category	Save and record supplier category data	Master_jenis_sup	Master	Disk1
Supplier	Save and record supplier	Master_supplier	Master	Disk1
Agricultural Product	Save and record agricultural product	Master_produk_tani	Master	Disk1
Agricultural product price	Save and record daily agricultural product price	Trans_harga_prod	Transaction	Disk1

Data Entity	Description	Logical Data Component	Data Category	Physical Data Component
Agricultural Product Offer	Save and record supplier offering data of agricultural tools, materials, and equipment.	Trans_penawaran_prod_tani	Transaction	Disk2
Agricultural Product Demand	Save and record agricultural product demand	Trans_permintaan	Transaction	Disk2
Agricultural Product Offer Log	Save and record supplier offering log data	Log_penawaran_prod	Log	Disk3
Agricultural Product Demand Log	Save and record demand log data	Log_permintaan	Log	Disk3
Chart of Account	Save and record farmers chart of account	Master_kode_biaya	Master	Disk1
Budgeting	Save and record farmer's production budget	Trans_anggaran	Transaction	Disk2
Budgeting Detail	Save and record farmer's production detail budget	Trans_detail_anggaran	Transaction	Disk2
Realization	Save and record production cost realization	Trans_realisasi_biaya	Transaction	Disk2
Cost Realization Log	Save and record farmer's log activity on realization budget	Log_biaya_keluar	Log	Disk3
Services	Save and record services for Integrated Agriculture Information System	Master_serv	Master	Disk1
Services Request	Save and record request for Integrated Agriculture Information System Services	Trans_permintaan_serv	Transaction	Disk2
Services Activity	Save and record log services activity	Log_activity_serv	Log	Disk3
Training Materials	Save and record training materials	Master_materi_train	Master	Disk1
Training request	Save and record training request from farmers community	Trans_permintaan_train	Transaction	Disk2
Fasilitator	Save and record faslitator detail data	Master_fasilitator	Master	Disk1
Training administration	Save and record training administration data	Trans_pelaksanaan_pelatihan	Transaction	Disk2
Training activity	Save and record log training activity data	Log_aktivitas_train	Log	Disk3
Learning Material	Save and record learning materials data	Master_materi_ajar	Master	Disk1
Learning Material Detail	Save and record learning materials data detail	Master_detail_materi_ajar	Master	Disk1

From table 1, there are 47 tables are involved for Integrated Agricultural Information System. Those 47 data entities are derived based on the information needed by farmers and stakeholder in agriculture sector. Those 47 tables are involved and related to 9 information systems in Integrated Agricultural Information System.

Class Diagram

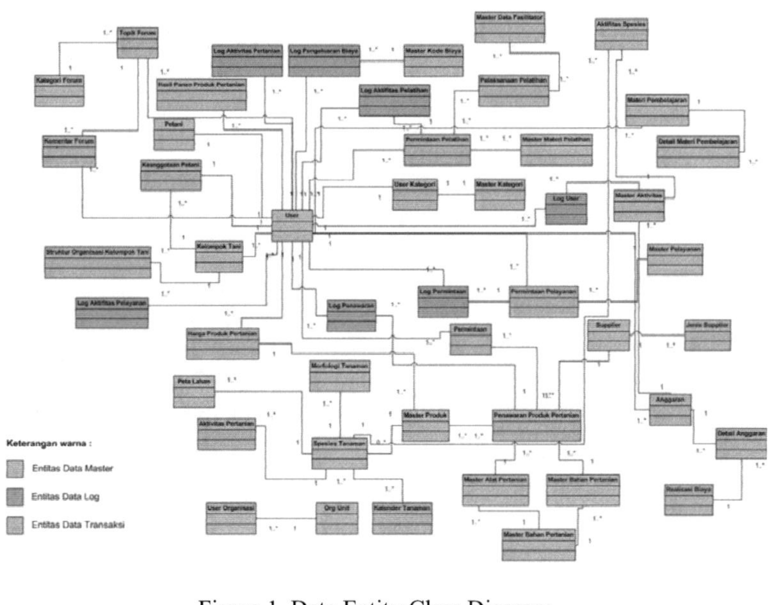

Figure 1. Data Entity Class Diagram

Figure 1 shows the class diagram of the data entities. There are 3 kinds of data entity: (1) Master Data Entity (Colour: Pink); (2) Log Data Entity (Colour: Blue); and (3) Transaction Data Entity (Colour: Green). There are 9 information systems are involved in Integrated Agriculture Information System: (1)

Agriculture Integrated Portal; (2) Farmer and Farmer Community Information System; (3) Farmers and Production Activity Information System; (4) Cropping Calendar Information System; (5) Agriculture e-Commerce; (6) Production Budgeting; (7) Service Information Systems; (8) Training Management System; (9) Learning Management System for Farmers.

4.2. System and Data Matrix

The purpose of the creating matrix of System and data is to depict the relationship between applications and the data entities that are accessed and updated by them. The mapping of the System and Data Entity relationship is an important step as it enables to assign access of data to specific applications in the organization, understand the degree of data duplication within different applications, and the scale of the data lifecycle, understand where the same data is updated by different applications, and support the gap analysis and determine whether any of the applications are missing and as a result need to be created. System and Data Matrix for IAIS can be seen in table 2.

Table 2. System and Data Matrix for IAIS

Application	Application Description	Data Entity
Agriculture Integrated Portal	This system is developed as a portal for integrated agriculture information system. This system also to fulfill farmer required innovation in agriculture processing and technique.	master_kategori, master_user, master_user_kat, master_user_org, master_org_unit, master_kategori_topik, Trans_topik_diskusi, trans_komentar_diskusi, Log_user_trans
Farmer and Farmer Community Information System	This system is developed as a data center for farmers and farmer's membership in community. This system is to help farmer community in order to organize its member and organization. This system is the foundation to build other system.	Master_user, master_kel_tani, master_petani, Trans_struk_org, trans_ang_petani
Farmers and Production Activity Information System	This system is developed to save production activity, such as preprocessing, land processing, harvesting, and post-harvest. This system also helps the government to monitor what farmers are doing. Farmers are able to get also some new technique and practise from expert	Master_user, master_petani_master_kel_tani, master_morf_tanaman, master_spesies_tanaman, master_aktifitas, master_aktifitas_spesies, Trans_ang_petani, trans_aktifitas_pertanian, trans_hasil_panen
Cropping Calendar Information System	This system is developed to administer cropping calendar from each region	Master user, master_user_org, Trans_kalender_tanam
Agriculture e-Commerce	This system is daveloped a gateway for farmers and business sector (private sector) can meet and sell their product. This system is to help farmer to get information about price and supply for product, tools, and materials.	Master_user, master_petani, master_kategori_alatbahan, master_alat_tani, master_bahan_tani, master_jenis_sup, master_supplier, master_prod_tani, Trans_hasil_panen, trans_harga_prod, trans_penawaran_prod_tani, trans_permintaan, Log_penawaran_prod, log_permintaan
Production Budgeting	This system is developed to help farmers manage their production cost.	Master_user, master_petani, master_aktifitas, master_aktifitas_spesies, master_kode_biaya, Trans_aktifitas_pertanian, Trans_anggaran, trans_detail_anggaran, trans_realisasi_biaya, log_biaya_keluar
Service Information System	This system is developed to help farmers get information and services from Integrated Information System, such as information on pest management, innovation in agriculture processing. This system also helps farmers meet the expert.	Master user, master_pelayanan, Trans_permintaan_pelayanan, Log_aktifitas_pelayanan
Training Management System	This system is developed to help farmer community to arrange training for its member. System will help to find the appropiate facilitator based on their expertise and research interest.	Master_user, master_fasilitator, master_materi_pelatihan,master_materi_ajar, master_detail_materi_ajar, Trans_permintaan_pelatihan, trans_pelaksanaan_pelatihan, Log_aktifitas_pelatihan
Learning Management System for Farmers	This system is designed as a tool for farmers to learn about technique and processing practise. It is also help farmers to get information about pest management, new products and techniques. This system is designed for farmers as a tool to learn.	Master_user, master_fasilitator, master_materi_pelatihan,master_materi_ajar, master_detail_materi_ajar, Trans_permintaan_pelatihan, trans_pelaksanaan_pelatihan, Log_aktifitas_pelatihan

Table 2 gives the detail system and its data entity for the system. There are 9 systems are involved and used the following data entity. This system will be developed gradually. The first phase is to build the Agriculture Integrated Portal, Farmer and Farmer Community Information System, and Farmers and Production Activity Information System. Implementation of the system build in the first phase will be followed by farmers training and empowerment to use the system.

IOP Conf. Series: Materials Science and Engineering **325** (2018) 012016 doi:10.1088/1757-899X/325/1/012016

5. Conclusion

Based on the analysis, it can be concluded that:

1. IAIS data architecture contains 47 data entities which are categorized into three: (1) Master Data; (2) Transaction data; and (3) Log data.
2. There will be 9 systems are involved in Integrated Agriculture Information System. Those systems are: (1) Agriculture Integrated Portal; (2) Farmer and Farmer Community Information System; (3) Farmers and Production Activity Information System; (4) Cropping Calendar Information System; (5) Agriculture e-Commerce; (6) Production Budgeting; (7) Service Information System; (8) Training Management System; and (9) Learning Management System for Farmer. In the first phase, it will develop agriculture integrated portal, farmer and farmer community information system, and farmers and production activity information system.

References

[1] Brugger F 2011 *Mobile Application in Agriculture* (Basen: Syngenta Foundation).

[2] Delima R 2016 *Analisis Kondisi dan Kesiapan Masyarakat Tani di Daerah Istimewa Yogyakarta untuk Memanfaatkan TIK di Bidang Pertanian* (Yogyakarta: Universitas Kristen Duta Wacana).

[3] Delima R, and Purwadi J 2015 Analisis Situs Web Pertanian Berbahasa Indonesia *Prosiding Seminar Nasional Komputer dan Informatika Terapan* pp 1–5.

[4] Delima R, Santoso H B, and Purwadi J 2016 Architecture Vision for Indonesian Integrated Agriculture Information Systems Using TOGAF Framework *Int. Conf. on Informatics and Computing* (Lombok: APTIKOM).

[5] Delima R, Santoso H B, and Purwadi J 2017 Business Architecture Development for Integrated Agriculture Information System (IAIS) Using TOGAF Framework. *Researchers World* VIII **2(1)** pp 1-13.

[6] Dunn C, and Grabski S 2001 An investigation of localization as an Element of Cognitive Fit in Accounting Model Representations *Decision Sci.* pp 55-94.

[7] Hoven J v 2003 Data Architecture: Principle for Data *Information Sys. Management* pp 93-6.

[8] Hoven J v 2004 Data Architecture Standards for the Effective Enterprise *Information Sys. Management* pp 61-4.

[9] Hutabarat T W 2012 *Mirisnya Menjadi Negara Pengimpor* ed K. Lumbanradja Retrieved Januari 7, 2017, from Mirisnya Menjadi Negara Pengimpor: http://blog-berbagi.blogspot.com/2012/07/indonesia-negara-agraris-omdo.html

[10] Kamran M, Anjum M, Rehman M, Ahmad H, and Kamran M A 2016 Classification of Information Systems in e-Agriculture: A Mapping Study *Int. J. of Comp. Sci. and Information Security* **14(9)** pp 1043-77.

[11] Kementrian Riset, Teknologi, dan Pendidikan Tinggi 2016 *Rencana Induk Riset Nasional 2015-2045.*

[12] Masiello-Riome C, Heller N, Rudgard S, and Schneider R 2008 Analysis of e-Agriculture Survey *Agricultural Information Worldwide* **1(1)** pp 11-8.

[13] Rees D, Momanyi M, Wekundah J, Ndungu F, and Odondi J 2000 Agricultural Knowledge and Information Systems in Kenya - Implication for Technology Dissemination and Development *Agricultural Research and Extension Network* (London).

[14] Renko N, Nikalosevic S, and Pavicic J 2002 The Market Information System and State Support for the Market of Agricultural Products in Croatia *British Food J.* **104(7)** pp 543-71.

[15] Santoso H B, and Delima R 2016. Stakeholder Definition for Indonesian Integrated Agriculture Information System (IAIS) *The Int. Conf. on Information Tech. and Digital Applications* (Yogyakarta: Universitas Islam Indonesia) pp 103-9.

[16] The Open Group 2011 *Architecture Development Method* [Online] Available: http://pubs.opengroup.org/architecture/togaf9-doc/arch/chap10.html

[17] Wen G, Zetian F, Daoliang L, Longyong Y, Jian X, and Xiashuan Z 2007 AgriInfo: an Agricultural Information System Based on a Call Center in China *New Zealand J. of Agricultural Research* **50** 797 – 806.

Acknowledgments
Special thanks to Ministry of Research, Technology, and Higher Education who give grant to fund our research. Researchers also thanks to Duta Wacana Christian University as our base who provides facility in doing this research.

When More is Less: The Case of Disconnected Information Systems in Indonesian Public Health Facilities

Fathul Wahid, Raden Teduh Dirgahayu, Almed Hamzah, Hari Setiaji

Department of Informatics, Universitas Islam Indonesia

{fathul.wahid, teduh.dirgahayu, almed.hamzah, heri.setiaji}@uii.ac.id

Abstract. The clear majority of previous studies have found that the absence of information systems to properly manage data is one of the main challenges in improving public health management. The present study offers an alternate perspective, revealing other emerging problems in cases where there are many information systems in place but without sufficient orchestration. The national government of Indonesia has been coercive in its implementation of various information systems without involving users at public health facilities, which has created many problems on the ground. The problems identified relate to the quality of the disconnected information systems currently in use, the lack of human resource development, unclear procedures, uncoordinated reports and the absence of an incentive scheme. The present study also highlights some practical implications, including the use of a more holistic perspective in designing and developing an integrated public health information infrastructure.

Keywords: system integration, silo thinking, health information system, information infrastructure, Indonesia, developing countries.

1. Introduction

Several previous studies have reported several challenges in managing public health in developing countries, and Indonesia is no exception. These challenges include weak healthcare systems and under-resourced facilities (Cline & Luiz, 2013). The use of information systems in this sector has been advocated, as they can help improve the quality of healthcare services and policies (Blaya, Fraser, & Holt, 2010; Cline & Luiz, 2013; Pappaioanou et al., 2003). From the literature, we can see that the absence of properly managed data contributes to poor policy formulation in the health sector (Pappaioanou et al., 2003).

The present study focuses on contexts in which various information systems are in place. The intriguing question regarding such cases is: Does the existence of information systems really contribute to better quality public health management? As measuring quality is not straightforward, we instead seek to answer the following question in this study: How are information systems used and managed in Indonesian public health facilities (called Pusat Kesehatan Masyarakat, or Puskesmas)? Answering this question will also provide insights to the former question. To achieve this, we conducted an interpretive case study involving several public health facilities in Indonesia.

The remainder of the paper is structured as follows. The second section provides a concise review of literature on the area of concern (i.e. the use of information systems in managing public health). In the third section, we describe the research context and methods. We present the research findings in the fourth section, which are then discussed in the fifth section. The paper ends with conclusions and recommendations.

2. Literature Review

From previous studies, we can observe that the application of information technology (IT) in healthcare varies, from personal use to institutional purposes. The present study focuses on the latter. An example of a system for personal use is the 'assisted living support system', designed to help the elderly during the medication period (Šabanović, Chang, Bennett, Piatt, & Hakken, 2015), to help patients confirm their diagnoses (Owens, 2015), or to support health education (Harpe, Town, & Africa, 2012). Meanwhile, the use of IT in institutions covers a variety of purposes, including to improve efficiency in data-sharing among the stakeholders (Woźniak, Romanowski, Yantaç, & Fjeld, 2014) and to support communication (Bowen, Dearden, Wright, Wolstenholme, & Cobb, 2010).

More specifically, the present study focuses on the use of data to support policy modelling, which has been deemed very important to improving the quality of decision making in public health (Pappaioanou et al., 2003; Sutcliffe & Court, 2005). Our search into the literature revealed two complementary approaches in policy modelling: evidence-based (Sutcliffe & Court, 2005) and through 'eParticipation' involving a group of people (Macintosh, 2004). The former uses evidence from the field to formulate policies, while the latter invites citizens to participate by providing their opinions.

There are several stages in policy modelling. The first stage is agenda setting, followed by policy construction, implementation and, finally, evaluation (Sutcliffe & Court, 2005; Young & Quinn, 2002). To formulate effective policies, in all stages, the interests of the stakeholders should be taken into thoughtful consideration. Nevertheless, many factors affect the quality of policies, in addition to the technical ones. These factors include the experience of the stakeholders, resource availability, organisation values, lobbying activities, pressures from certain groups, pragmatism, legal aspects, media pressure and even luck (Davies, 2004; Zardo, Collie, & Livingstone, 2014).

The present study is considered a preliminary step towards effective policy formulation, focusing on mapping the use of information systems in Indonesian public health facilities, which is needed for policy construction. In this case, an integrated system is a necessity (De Kadt, 1989); otherwise, the systems in place will be fragmented and overlapping, which is often what happens in some national contexts (Okuonzi & Macrae, 1995).

3. Research Method

The present research is an interpretive case study (Walsham, 1995). Stories and knowledge from the stakeholders comprise the main sources of data. To elicit this information, we interviewed five key actors across three public health facilities, lasting for four hours altogether. The interviewees are knowledgeable operators of information systems and have various backgrounds, including staff from the medical records, administrative, and nutritional departments. The data collection was carried out in early August 2017. This represents the preliminary data collection activities before a mass survey is conducted (in our future study).

We decided to analyse the qualitative data in a group discussion, involving the four authors of the present study, to gain further meaningful insights. Our accumulated knowledge is very beneficial in interpreting the data. One of us has extensive experience in developing and implementing information systems for hospitals in various regions. Another is strongly familiar with similar problems, especially related to the implementation of financial information systems in the public health facilities of Indonesia. Interpreting the data was not straightforward, instead requiring several rounds of analysis. Through this, new interpretations emerged and were then compiled to form a more complete picture.

4. Findings

Our analysis of the data offered many interesting and important insights, although they should be considered preliminary in nature. First, we found various information systems in place. One of our interviewees reported that every health facility operated 15 information systems. Most are supplied by

IOP Conf. Series: Materials Science and Engineering **325** (2018) 012017 doi:10.1088/1757-899X/325/1/012017

a national government agency, which, in this case, is the Ministry of Health. The problems experienced by the health facility now are primarily concerned with how to run all these information systems at the same time and with limited resources.

Second, in this context, the nonexistence of information systems is not the problem. Rather, the problem emerges when those systems are not well orchestrated. The systems do not help the staff accomplish their duties, and the system becomes a burden because the operators on the ground must enter the same data into more than one system. Data duplication, therefore, has become an issue.

Third, the problem could be becoming worse, as we found that public health facilities were never involved in the design and development of the information systems they use. This was obviously a top-down decision, and no attention was made to discern technical requirements from the operators' perspectives. For example, one interviewee reported that the health facility stopped using the Sistem Informasi Kesehatan Daerah (SIKDA, the Indonesian regional health information systems), since the system cannot accommodate policies made by the Badan Penyelenggara Jaminan Sosial (BPJS, the Healthcare and Social Security Agency of Indonesia), in which patients are exempt from all the expenses they may incur.

Fourth, it can be understood that various information systems demand capable human resources to manage them. Unfortunately, all our interviewees admitted that they face problems in this regard. Most of the operators have no training nor background in IT. The facilities worked around this by assigning the responsibilities of managing certain information systems to specific people. One of our interviewees stated, 'One staff member may be assigned one to two information systems to manage'.

Fifth, the situation depicted above led us to the conclusion that, in the field, the staff member who should be responsible for operating the information systems is unclear. We found that each health facility has only one staff member with an educational background in IT. Hence, it is not difficult to imagine that all the information systems are beyond that staff member's capacity to manage.

Sixth, our interviews revealed that various reports should be regularly submitted by the health facilities. The reports are related to, for example, financial statements, immunisations, pilgrimage (hajj), clinical patients and special reports (e.g. HIV incidents). However, some of the information systems cannot produce a ready-to-send report format. Instead, the assigned operator must compile the reports manually. Some reports are submitted in print, while others can be uploaded into a shared storage folder on Google Drive.

Seventh, although we found that the reports are verified, there is no standardised verification procedure. For example, one of our interviewees reported that the data contained in the submitted report can be revised by sending a message through a WhatsApp group. This hinders the traceability of data versions and exchange.

Finally, even though the assigned operators have additional duties in addition to their official ones, no incentive wages are paid to them. From the perspective of human resource management, this is not a good practice for ensuring sustainability and increasing staff morale.

5. Discussion

The findings presented reveal some significant issues that need further discussion. First, 'sectoral ego',[1] or silo thinking, is commonplace. This 'public secret' of the public-sector management in Indonesia has an extensive influence on the design and development of information systems. As a consequence, a series of disconnected information systems were produced with few changes made to properly integrate them. When the system was web-based, the introduction of a 'bridge', such as an application programming interface (API), could help better integrate the systems. To ensure the information systems provide optimal benefits in the future, increased cooperation in designing and developing the systems is required. One of our interviewees suggested that 'the systems should be crafted into a single system to provide more integrated data. However, this needs good infrastructure and qualified human resources.' Silo thinking has been found to be a great burden in many contexts (Bygstad, Hanseth, & Le, 2015).

[1]

Various solutions have been recommended, such as implementing process thinking, defining standards and interoperability, developing enterprise resource planning (ERP)-like solutions and building enterprise architecture (EA) (Bygstad et al., 2015).

Second, it seems that developers of new information systems always start from scratch, neglecting the existing foundation (i.e. information systems currently in use that could be further developed). This neglect has led to a missed opportunity in developing the plethora of information systems currently available into an information infrastructure (e.g. Star & Ruhleder, 1996; Wahid, 2013). In addition, the lack of an integrated system creates further problems, such as high-energy-consuming reporting systems. One interviewee said, 'The report format should be unified'. Another interviewee unhappily criticised, 'If we use the application, we should able to send the report immediately'.

6. Concluding Remarks

We have reported findings from the field about the use of information systems by public health facilities in Indonesia. Our impression is that, from the perspective of the system operators, there are many problems in the systems without any clear and well-designed solutions. These problems are related to the quality of the disconnected information systems currently in use, the lack of human resources, unclear procedures, uncoordinated reports and the absence of an incentive scheme.

These findings reveal some practical implications. A more holistic perspective is needed in the design and development of public health information systems. Silo thinking should also be minimised. Only through these methods can an integrated information infrastructure be initiated and developed. Equally important is the involvement of the operators on the ground, as they represent precious information sources for determining the technical requirements.

As this is a preliminary study, limitations are inherent. We will validate our findings by conducting a survey of around 100 health facilities located in various locations (e.g. urban, rural, etc.) to gain a more complete picture about the patterns of information-system use. The findings will provide further insights into the development of a better policy-modelling system.

References

[1] Blaya J A, Fraser H S F, and Holt B 2010 E-health technologies show promise in developing countries *Health Affairs* ***29*****(2)** pp 244–51.

[2] Bowen S, Dearden A, Wright P, Wolstenholme D, and Cobb M 2010 Participatory healthcare service design and innovation *Proc. of the 11th Biennial Participatory Design Conf. on - (PDC '10)* (New York, USA: ACM Press) p 155 http://doi.org/10.1145/1900441.1900464.

[3] Bygstad B, Hanseth O, and Le D T 2015 From IT silos to integrated solutions. a study in e-health complexity *Proc. of the European Conf. on Information Syst. (ECIS) 2015.*

[4] Cline G B, and Luiz, J M 2013 Information technology systems in public sector health facilities in developing countries: the case of South Africa *BMC Medical Informatics and Decision Making* ***13*****(1)** p 13.

[5] Davies P 2004 Is evidence-based government possible *Fourth Annual Campbell Collaboration Coll.* **19**.

[6] De Kadt E 1989 Making health policy management intersectoral: issues of information analysis and use in less developed countries *Social Sci. and Medicine* ***29*****(4)** 503–14.

[7] Harpe R D, Town C, and Africa S 2012 Lessons learnt from the participatory design of a mobile care data application in a resource-restricted context. *Pdc '12-2* pp 125–8. http://doi.org/10.1145/2348144.2348184

[8] Macintosh A 2004 Characterizing E-Participation in Policy-Making *0*(C) 50117a p 1-10 http://doi.org/10.1109/HICSS.2004.1265300

[9] Okuonzi S A, and Macrae J 1995 Whose policy is it anyway? International and national influences on health policy development in Uganda *Health Policy and Planning* **10(2)** pp 122–32.

[10] Owens O L 2015 Principles for developing digital health interventions for prostate cancer: a community-based design approach with african american men *Human Aspects of IT for the*

Aged Population. Design for Everyday Life (Springer) pp 134–45.
[11] Pappaioanou M et al 2003 Strengthening capacity in developing countries for evidence-based public health: the data for decision-making project *Social Sci. & Medicine* **57(10)** pp 1925–37.
[12] Šabanović S, Chang W L, Bennett C C, Piatt J A, and Hakken D 2015 A robot of my own: participatory design of socially assistive robots for independently living older adults diagnosed with depression *Human Aspects of IT for the Aged Population. Design for Aging* (Springer) pp 104–14.
[13] Star S L, and Ruhleder K 1996 Steps toward an ecology of infrastructure: Design and access for large information spaces *Information Syst. Research* **7(1)** pp 111–34.
[14] Sutcliffe S, and Court J 2005 *Evidence-Based Policymaking: What is it? How does it work? What relevance for developing countries?* (London: Overseas Development Institute).
[15] Wahid F 2013A triple-helix model of sustainable government information infrastructure: Case study of the eProcurement system in the Indonesian public sector *Building Sustainable Information Systems* ed H Linger et al (Berlin: Springer) pp 555–67.
[16] Walsham G 1995 Interpretive case studies in IS research: nature and method *European J. of Information Syst.* **4(2)** pp 74–81 http://doi.org/10.1057/ejis.1995.9
[17] Woźniak P, Romanowski A, Yantaç A E, and Fjeld M 2014 Notes from the front lines: Lessons learnt from designing for improving medical imaging data sharing *Proc. of the 8th Nordic Conf. on Human-Computer Interaction Fun, Fast, Foundational (NordiCHI '14)* pp 381–90. http://doi.org/10.1145/2639189.2639256
[18] Young E and Quinn L 2002 *Writing Effective Public Policy Papers*. (Budapest: Open Society Institute).
[19] Zardo P, Collie A, and Livingstone C 2014 External factors affecting decision-making and use of evidence in an Australian public health policy environment *Social Sci. and Medicine* **108** pp 120–7. http://doi.org/10.1016/j.socscimed.2014.02.046

Vulnerability in Determining the Cost of Information System Project to Avoid Loses

Kholid Haryono[1] and Zulfa Amalia Ikhsani

Department of Informatics, Faculty of Industrial Technology, Universitas Islam Indonesia, Jl. Kaliurang KM. 14.5 Sleman Yogyakarta Indonesia

[1] Corresponding author: kholid.haryono@uii.ac.id

Abstract. Context: This study discusses the priority of cost required in software development projects. Objectives: To show the costing models, the variables involved, and how practitioners assess and decide the priorities of each variable. To strengthen the information, each variable also confirmed the risk if ignored. Method: The method is done by two approaches. First, systematic literature reviews to find the models and variables used to decide the cost of software development. Second, confirm and take judgments about the level of importance and risk of each variable to the software developer. Result: Obtained about 54 variables that appear on the 10 models discussed. The variables are categorized into 15 groups based on the similarity of meaning. Each group becomes a variable. Confirmation results with practitioners on the level of importance and risk. It shown there are two variables that are considered very important and high risk if ignored. That is duration and effort. Conclusion: The relationship of variable rates between the results of literature studies and confirmation of practitioners contributes to the use of software business actors in considering project cost variables.

1. Introduction

One of the success parameters software development project is the economic aspect in which the costs incurred can be controlled and the revenue is in line with expectations [1]. The difference is also called profit or loss. Unfortunately, there are still many software developers hit by iceberg phenomenon where the problem seems small when in fact the problem is big, broad, and deep. As a result the costs can be out of control and be among the causes of project failure.

Knowledge of the costing factor to control it is important because most software developers cannot quickly estimate the costs involved [2]. The hasty question is often asked by the user is what the price of software offered. Answering these questions require knowledge of the costs involved in software development. This is needed so that practitioners can use it to determine prices based on the cost variables involved. These variables have been shown in various models of cost estimation that have been widely used [2]

This study focuses on four issues, namely (a) what model is used to estimate software development costs; (b) what factors are taken in each model; (c) categorize the factors taken and prioritization of each category; and (d) how software developers perceive each category and provide a risk assessment of neglected categories. The purpose of this study is to show the factors that influence in determining cost of software, comparison between literature and assessment of practitioners, and to determine the level of risk of these variables that could potentially cause project losses.

The results provide new insights and knowledge for stakeholders in using the advantages and disadvantages of a factor in determining project costs. This knowledge is expected to improve the quality of decisions related the project costing, thus avoiding project losses. The quality of decisions taken by software business practitioners about efforts to reduce project costs to achieve project success. To support the knowledge in the decision-making, any factors found will be confirmed. There are two confirmed, i.e, important level and risk level to software developers.

2. Motivation

We have two motivations for this research. First, provide important information about the factors or variables that affect the cost of software development for the perpetrators. Second, it gives confirmation of the influencing factors to the software business practitioners mainly related to the cost and the risks.

3. Research methodology

This study focuses on three stages, namely (a) literature review to look for software costing models. Each model is deepened by looking for its cost variables. The most variables appearing on the model are given the highest priority and order to the lowest. (b) empirical variable collection, all variables are confirmed to IT business practitioners. Correspondents are asked to assess the importance of each variable with the degree of importance 1 to 5. The higher the value is the more important the variable has an effect on the variable costing software. These variables are also confirmed about the level of risk posed if the variable is ignored. The risk rating of scale 1 to 5 also with value 5 has the greatest degree of risk (c) discusses the findings to discover new knowledge contributions:

3.1. Systematic Literature Review

Systematic Literature Review (LRS) has been widely used to observe, categorize, and evaluate various literature to reveal facts related to research areas undertaken by researchers by applying two techniques namely inclusion and exclusion[3]. Systematic literature can be used for various fields. Systematic literature in accordance with the field of software engineering has been reviewed by Kitchenhan[4]. Most researchers use three phases in LRS, that are planning, conducting, and reporting.

3.1.1. Planning the review.

Planning starts from determining the question for literature search. Search queries include: estimation model costing software development. The found models are then deepened by searching for the variables used each model with questions according to the model variables concerned.

The first source of data from general search through google. Some of the models found are LOC, COCOMO, COSYSMO[5], SLIM, PRICE-S, SEER-SEM[6, 7], PROBE[8], EBS[9], REVIC[10]. Each model is in the search for variables that are taken into account each variable.

Early models were found on Wikipedia with the keyword "Software development effort estimation" or "Cost estimation in software engineering". The models found are used as keywords to search for variables through scientific publications (journals and proceedings). This further study is used to determine the context of the variables used in each model. Selection of paper using inclusion and exclusion criteria to determine the appropriate paper. To make sure the paper is searched properly, keyword searching and paper collection are determined. The selected papers are selected through the title and abstract content

3.1.2. Conducting the review.

Based on data searches conducted using scientific data sources 2016 and 2017 with the filter "original research" obtained the number of papers such as **Table 1**.

Table 1. Source publication and amount found

No	Sources	Found	Relevant
1	Science Direct	131	19
2	Google Scholar	160	11
Amount		291	30

3.1.3. Reporting the review.

Based on the literature, the variables of each model are shown in **Table 2**.

Table 2. The variables in model

No	Model Name		The variables involved
1.	COCOMO	:	Staffing; Productivity; Effort; Duration; Maintenance
2.	COSYSMO	:	Personal Team; Size; Maintenance; Migration; Complexity; Process Business; Stakeholder Cohesion; Risk; Documentation;Tool Support
3.	REVIC	:	Staffing; Productivity; Effort; Duration; Maintenance;Integration; Testing Complexity
4.	SEER-SEM	:	Staffing; Productivity; Effort; Duration
5.	SLIM-Putnam Model	:	Productivity; Size; Effort; Duration; Function
6.	PRICE-S	:	Scope; Effort; Scheduling; Cost
7.	Evidence Based Scheduling	:	Time; Activity; Budget
8.	ITK Method-CETIN	:	Labor maintenance; Labor middleware; Function Size; Middleware; Complexity; Design; Budget
9.	PROBE	:	Size; Time; Quality; Effort
10.	Software Equation	:	Productivity; Effort; Line of Code; Skill Factor; Time

Furthermore, the variables of each model are grouped by their similarity. For example staffing variables are owned by five models namely COCOMO, COSYSMO, REVIC, SEER-SEM, and CETIN[11]. The variables arranged according to the priority are shown in **Table 3.**

Table 3. List of variables from each group

No	Variabel		Amount Model	Term on the model
1.	Duration	:	8	Duration; time; scheduling
2.	Effort	:	7	Effort
3.	Complexity	:	7	Complexity to understanding; Process business capability; Testing complexity; Skill factor; Design complexity; Stakeholder cohesion
4.	Staffing	:	6	Staffing; Personal team; Labor
5.	Size	:	5	Size; Scope; Function Size
6.	Productivity	:	5	Productivity platform; Productivity
7.	Budget	:	3	Cost; Budget
8.	Maintenance	:	3	Maintenance
9.	Function	:	2	Function; Task activity
10.	Integration	:	2	Integration; Migration
11.	Quality	:	1	Quality
12.	Midleware	:	1	Midleware
13.	Line of Code	:	1	LOC
14.	Tools Support	:	1	Tools Support
15.	Documentation	:	1	Documentation

3.2. Empirical data collection

This stage confirms the variables found in the literature study to software business actors. There are two confirmed, which are related to the degree of importance and the degree of risk that can occur if those variables are ignored.

Findings on the degree of importance will be compared with the priorities found in the literature study. This comparison becomes important information especially if there is a significant difference because the difference can be new knowledge that is very useful for decision-making in the future. Risk assessment can be used as knowledge of potential losses and strong suggestions on the use of variables in project costing

Wherever possible try to ensure that the size of the text in your figures (apart from superscripts/subscripts) is approximately the same size as the main text (11 points).

3.2.1. Respondents.

Conditions of respondents in this study there are two, the first person who is active in the software company, and second on the company, this person has a position to calculate the cost of production of IT projects and or determine the selling price of the software. Based on the distribution of questionnaires, the participating respondents came from 14 companies with the number of respondents 15. The names of the companies are: CV.Ainan; Partikelir; Prima solusi hexa; PT. Krafthaus Indonesia; Maylab; Adhi jasa informatika; CV. SAK; Ngrembel Asri; PT Skyshi Digital Indonesia; Smallstep Labs; Folarium; Orion Techno; Softwareseni; and ATSoft.

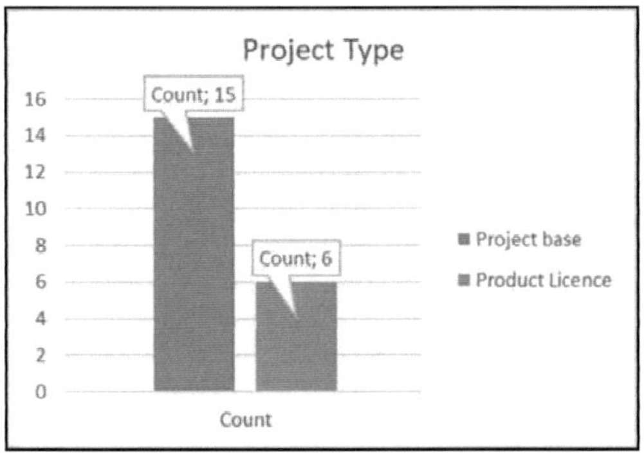

Figure 1. The product type of these companies

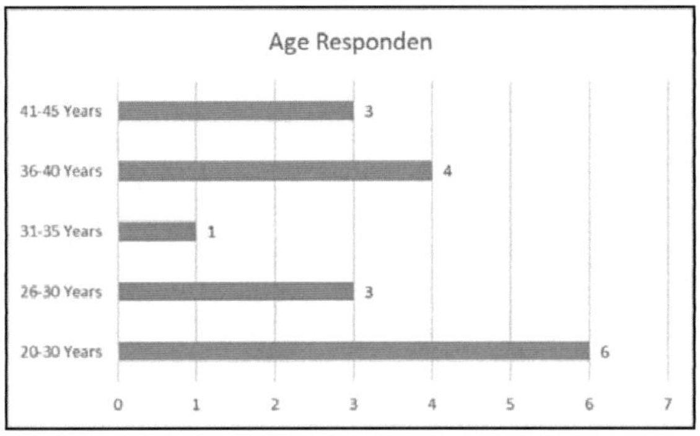

Figure 2. Age of respondents ranged from 22-43

The customer category of the company is divided into three, government (6), private (14), and abroad (4). Based on the educational background, the average of the last education of the respondent is S1. Distribution of educational background is 80% Bachelor degree, 13% Master, and 7% High school. The sex of the respondents is predominantly male. Of the 15 respondents, only one is female.

3.2.2. Survey and analysis.

The questionnaires that were distributed included two assessments namely the degree of importance and the degree of risk. The list of questions and number of respondents is shown in **Table 4**.

The variables are sorted according to degree of importance and degree of risk. To calculate the highest score is by multiplying each variable with weight (1-5) the result is divided by the number of respondents it will produce a score of 1 - 5. The highest score shows the high degree of importance. The graph of degree of importance and degree of risk is shown in **Figure 3**.

Comparison of importance and risk levels are at the same level. It shows the perception of respondents to each variable is in accordance with the meaning of the variable. Only middleware variables alone have too high a perception gap. This happens because middleware includes the unpopular in determining the cost of developing an IT project

Table 4. List of questions and amount answers respondents

No	Element of Project Costing	Degree of importance 5 is most important					Degree of risk 5 is most risky				
		5	4	3	2	1	5	4	3	2	1
V1	Duration	11	4	0	0	0	11	4	0	0	0
V2	Effort	8	7	0	0	0	8	6	0	1	0
V3	Complexity	3	10	1	1	0	4	6	3	2	0
V4	Staffing	5	5	4	1	0	6	6	3	0	0
V5	Size	3	8	2	2	0	2	7	2	4	0
V6	Productivity	2	7	4	2	0	1	7	4	3	0
V7	Budget	6	7	2	0	0	6	8	0	1	0
V8	Maintenance	4	7	2	2	0	7	6	1	1	0
V9	Function	8	5	1	1	0	9	4	0	2	0
V10	Integration	2	10	2	1	0	4	9	1	1	0
V11	Quality	7	6	2	0	0	7	6	2	0	0
V12	Middleware	3	4	5	2	1	6	6	2	1	0
V13	Line of Code	5	7	3	0	0	4	7	2	2	0
V14	Tool Support	7	7	1	0	0	8	5	1	1	0
V15	Documentation	4	6	4	1	0	4	7	3	1	0

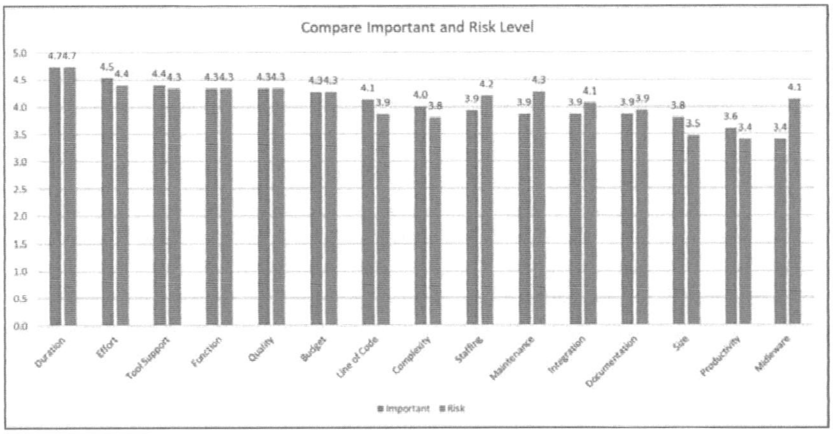

Figure 3. Graph of of degree of importance and degree of risk

4. Finding

Based on the data obtained from the various models and empirical data, found a shift in the variable priority between the model, importance level, and risk level of each variable. The shift is shown in **Figure 4**.

The findings show two consistent variables occupying the topmost level. Namely, duration and effort. This shows that the highest priority in determining cost of software development is the length of time required and the amount of effort that must be prepared and done.

Complexity variables dropped the 3rd priority position into the 8th importance and the level of risk further dropped to the order of 13. This means, according to the respondents varied complexity is on the middle priority in considering the cost of software development. Respondents assessed the risk level for this variable is also low if ignored so that this risk level data reinforce the decline in importance level.

Important Variable	Important level
Duration	4.7
Effort	4.5
Tool Support	4.4
Function	4.3
Quality	4.3
Budget	4.3
Line of Code	4.1
Complexity	4.0
Staffing	3.9
Maintenance	3.9
Integration	3.9
Documentation	3.9
Size	3.8
Productivity	3.6
Midleware	3.4

No	Variable	Count in models
1	Duration	8
2	Effort	7
3	Complexity	7
4	Staffing	6
5	Size	5
6	Productivity	5
7	Budget	3
8	Maintenance	3
9	Function	2
10	Integration	2
11	Quality	1
12	Midleware	1
13	Line of Code	1
14	Tools Support	1
15	Documentation	1

Risk Variable	Risk level
Duration	4.7
Effort	4.4
Function	4.3
Quality	4.3
Tool Support	4.3
Budget	4.3
Maintenance	4.3
Staffing	4.2
Midleware	4.1
Integration	4.1
Documentation	3.9
Line of Code	3.9
Complexity	3.8
Size	3.5
Productivity	3.4

Figure 4. The shift variables between importance level, risk level, and literature urgency

In contrast, the variable quality rises from the order of the 11th model to the 5th importance level with the 4th risk level. This shows that quality is an important priority that has a high enough risk if ignored. Similarly, other variables.

5. Discussion

A discerning look at the variables involved in determining the cost of a software product becomes important because it involves the success of a project. In business, the success of a project is determined by the amount of the difference in cost and income received. Failure to see these variables can cause losses due to costs that should not be issued.

Based on empirical data collection, the priority level of the variable interest can be divided into three parts. First, the top priority variables of order 1-5 are duration, effort, tool support, function, and quality. This section is the top priority that is considered important in determining the cost of software. This section is supported by the highest risk level assessment. Rankings 1-5 on risk assessment are also occupied by these variables although the sequence is different.

Second is the five medium priorities. They are budget, LoC, complexity, staffing, and maintenance. Budget is client budgeted cost or client purchasing power to software project. Line of Code (LoC) is the number of program lines each function. Complexity is the uncertainty of problems and business procedures that will be resolved. A small visible problem shows the complexity of a project because the real problem will be found in the course of project work. Staffing is a team requirement, including experts who need to be hired. Maintenance is to consider the maturity level of the company/client. The less maturity IT will be the greater the cost of maintenance budgeted because the business is still full of uncertainty.

Budget, staffing, and maintenance at this intermediate level, risk level variable according to the variable of importance level. The other two variables are middleware and integration that outperform LoC and complexity variables. Middleware is a service that bridges between software. This section is strongly associated with integration so that two of these by the respondents considered quite risky.

Third, the variables that occupy the lower interest level are integration, documentation, size, productivity, and middleware. Although in the lower group, the average importance of this level is 3.7 on a scale of 1 to 5. The same lower risk level is documentation, size, and productivity. Documentation is generally attached to the record of each process so that it becomes the usual priority. Size is the project boundary. The size limitations are difficult to measure for current problems due to the rapidly changing developments and competition. While productivity is more on the ability of the team formed to complete the project of each feature.

6. Future work

The findings discussed in this study are important to be tested directly on the practice of software development projects. The order of variables can be a guide to the determination of weights on costing. The composition of weights can be tailored to the project and the policies used. Priority risk variables can also be used as a calculation of risk management weights in software development. The costing of this paper is more directed to software development companies based on orders. This is to be tailored to the needs of the company. For cloud computing company with subscription system is not covered in this research.

7. Conclusion

The models that can be used in determining the cost of software development are at least 10 models as shown in Table 3. Each model has variables used for calculation. The total variables collected were 54. The variables are grouped/categorized based on similarity of meaning. The result is 15 main groups called 15 main variables. These variables are shown in Table 5. The variables have been sorted by the number of variable members in the model. The more models that have a particular variable then based on the higher the priority of the variabel. The emerging variables are confirmed to the software business through questionnaires. There are two confirmations, namely the level of importance and the level of risk. Two consistent variables at the top are duration and effort. Both have the highest level of importance and are considered very risky if ignored. The sequence of variables can be used as an important consideration in calculating the cost of software development in practice.

References

[1] Niazi M, Mahmood S, Alshayeb M, and Majid A 2016 Toward successful project management in global software development *JPMA* **34** pp 1553–67.
[2] Potdar S M, Ingle A, Puri M, and Potdar M 2014 Factors influencing on cost estimation *Glob. J. Adv. Eng. Techologies* **3** pp 119–23.
[3] Barbara K *et al* 2009 Systematic literature reviews in software engineering – A systematic literature review *Inf. Softw. Technol.* **51** pp 7–15.
[4] Kitchenham B 2004 *Procedures for Performing Systematic Reviews*.
[5] Živadinović J, Medić Z, Maksimović D, Damnjanović A, and Vujčić S 2011 Methods of effort estimation in software *Int. Symp. Eng. Management And Competitiveness* **2011** pp 417–22.
[6] Ziauddin S, Kamal T, and Ziauddin Z 2012 An effort estimation model for agile software development *Adv. Comput. Sci. its Appl.* **2** pp 314–24.
[7] Rajper S and Shaikh Z A 2016 Software Development cost estimation : a survey *Indian J. Sci. Technol.* **9** pp 1–5.
[8] Besimsee W B 1996 *The PSP Size Estimating Approach* **1**.
[9] Spolsky A J 2008 *More Joel on Software* 320 (Apress). doi:10.1007/978-1-4302-0988-1
[10] Strickland D 1999 REVIC 11 : Converting the REVIC Model to COCOMO I1 *Center for Systems and Software Eng.* **35807**.
[11] Khatibi V and Jawawi D N A 2010 Software Cost estimation methods : a review *J. Emerg. Trends Comput. Inf. Sci.* **2** pp 21–9.

Acknowledgments

Our thanks to Informatics Department of Universitas Islam Indonesia for allowing us to use facilities, academic atmosphere, and funding support in this study.

IOP Conf. Series: Materials Science and Engineering **325** (2018) 012019 doi:10.1088/1757-899X/325/1/012019

Characterization of Soil Moisture Level for Rice and Maize Crops using GSM Shield and Arduino Microcontroller

G A Gines[1], J G Bea[2] and T D Palaoag[3]

[1]Lorma Colleges, City of San Fernando, La Union, Philippines

[2]Ilocos Sur Polytechnic State College, Tagudin, Ilocos Sur, Philippines

[3]University of the Cordilleras, Baguio City, Philippines

ggines@lorma.edu, joygraycochea@yahoo.com, tpalaoag@gmail.com

Abstract. Soil serves a medium for plants growth. One factor that affects soil moisture is drought. Drought has been a major cause of agricultural disaster. Agricultural drought is said to occur when soil moisture is insufficient to meet crop water requirements, resulting in yield losses. In this research, it aimed to characterize soil moisture level for Rice and Maize Crops using Arduino and applying fuzzy logic. System architecture for soil moisture sensor and water pump were the basis in developing the equipment. The data gathered was characterized by applying fuzzy logic. Based on the results, applying fuzzy logic in validating the characterization of soil moisture level for Rice and Maize crops is accurate as attested by the experts. This will help the farmers in monitoring the soil moisture level of the Rice and Maize crops.

1. Introduction

Soil serves as a medium for plants growth. It absorbs water, keeps moisture and holds the roots of the plants. To irrigate crops, one method commonly used is to follow soil moisture depletion. As a plant grows, it uses the water within the soil profile of its root zone. As the water is being used by the plants, the moisture in the soil reaches a level at which irrigation is required or the plant will experience stress. If water is not applied, the plant will continue to use what little water is left until it finally uses all of the available water in the soil and dies [1]. One factor that affects the soil moisture is drought. Drought has been a major cause of agricultural disaster [2]. Agricultural drought is said to occur when soil moisture is insufficient to meet crop water requirements, resulting in yield losses [3].

To avoid yield losses, a device was developed in India to detect the soil moisture level. The device is placed in the field and works under three conditions wet, normal and dry conditions. If Wet and Normal Condition the device remains constant. If the soil moisture is Dry condition, then the water pump is turned on until it reaches the normal condition [4]. Wireless sensor network is another technology developed in India. It is use in automated irrigation system to enhance the use of water for agricultural crops. It is consists of a soil moisture sensor and temperature sensor placed under the soil where plants roots are reached [5].

Drought is affecting also the large tracts of the main Rice-producing areas of Asia. The Ministry of Environment in Cambodia estimated that between 1996 and 2000 drought-related Rice losses were approximately 20 percent of potential national production. Severe drought in 2004 affected 300,000 hectares of paddy Rice leading to an 82 percent loss of the potential harvest. And in 2009 another severe drought was recorded and the government issued a USD 12 million rescue package [6]. In the United States, export pRices for corn soared nearly 128 percent above the 20-year historical average. Export pRices also hit the highest level since the import and export pRice index series began in December 1984.

Further, In North Dacota, they used "check-book method to solve the problems in reduced yields. This method uses daily crop water use values and soil water holding capacities to predict the time and amount of water needed to replenish what has been removed from the root zone. Rain and irrigation amounts are deposits for storage in the root zone, and crop water use is a withdrawal of water from that storage [7].

Moreover, the farmers in the Philippines await the coming of rains before planting the season's crop. A good cropping is highly dependent on sustained rainfall, especially during the critical stages of crop development. The farmers water their plants using irrigation and water pump. Most farmers used "Counting Method" to determine when to water the plants. This process sometimes is not accurate because of the climatic irregularities could spell disaster [8] which leads to crop failure.

With the above scenario, there is a felt need to integrate technology in determining the soil moisture level to improve the Rice and corn production. This will also help the farmers to manage their crop properly and to avoid loss of crop production. Thus, the objective of the study is to characterize the level of soil moisture level for Rice and Maize crops using Arduino with Fuzzy Logic. Arduino Mega is a microcontroller which has a number of facilities that communicates with other microcontroller, sensors and computer. Fuzzy Logic Algorithm was used to characterize the values gathered from the Arduino Mega.

2. System Architecture

Figure 1 shows the Arduino Mega as the main controller of the entire system. All other components are directly connected to the controller. The soil moisture sensor rod is directly planted to the soil as the i2c is connected to the controller. The sensor detects the moisture level of the soil and gives the information to the controller. A text message containing the soil moisture value and the action to be taken will be send immediately to the farmer thru GSM Shield. The sensor is also capable of switching to different crop such as Rice and Maize using toggle switch. The GSM Shield will carry all SMS send and receive by the farmer. The solar panel serves as the power source of the different components.

Figure 2 shows the System Architecture for Water Pump. The Water Pump is connected to the Arduino Mega via switching relay as plugged to 220v AC. When the soil gets dry, a SMS will be sent automatically to the farmer and the farmer has the option to turn on or off by sending SMS "ON" or "OFF", respectively. The switching relay is triggered when the farmer send SMS to the GSM Shield to turn on the water pump as the water pump already plug to the AC.

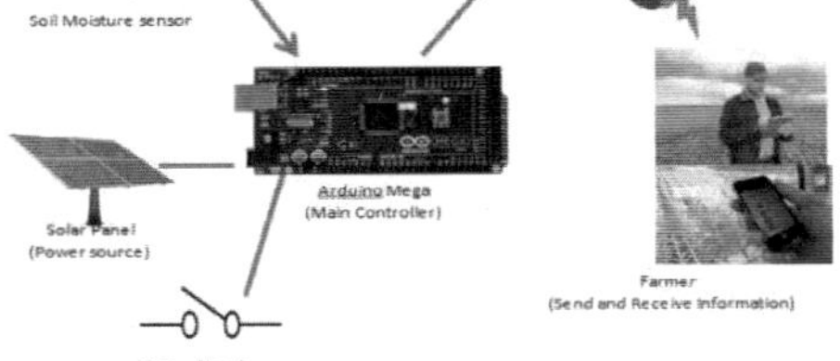

Figure 1. System Architecture for Soil Moisture Sensor

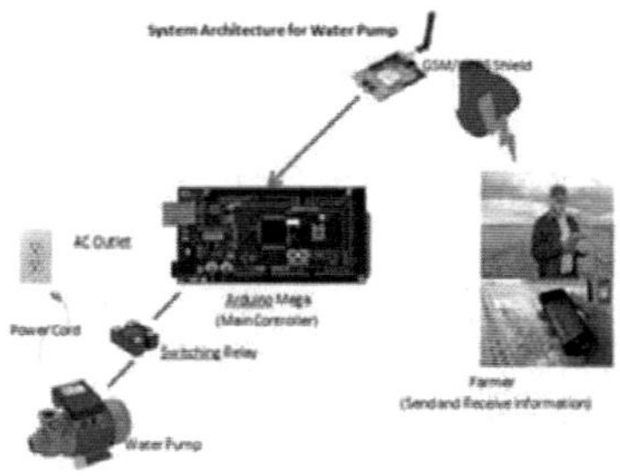

Figure 2. System Architecture for Water Pump

3. Methodology
This study used descriptive and developmental type of research design. Descriptive research is significant as surveys abound in educational research and are utilized by many researchers as an investigative tool to collect data in order to address educational questions [9]. Developmental research design seeks to create knowledge grounded in data systematically derived from practice. It is a way to establish new procedures, techniques, and tools based upon a methodical analysis of specific cases [10].

The researchers designed system architecture for soil moisture sensor and water pump. Fuzzy logic was integrated in characterizing soil moisture level for Rice and Maize crops. As proven in the study of [11], fuzzy logic has been used in: process control, management and decision making, operations research, economics and engineering. Fuzzy logic is more stable to use and estimation of soil moisture is more balance, thus, this study will be using fuzzy logic for characterizing soil moisture level for Rice and Maize.

Purposive sampling was used in determining the experts. The experts who were available at the testing period validated the characterization of the soil moisture value detected by the soil moisture sensor. For example, for Rice crops soil moisture value, if the device detected 270, the experts characterized it as Wet. This means that the soil moisture level of the Rice crops is still wet and do not need to be watered. And if the device detected 435, the experts characterized it as Dry. This means that the soil moisture level is dry so it needs to be watered.

3.1. Hardware Components
The following hardware components below were used in the study.
- Arduino Mega - The Arduino Mega is the main controller of all the components in the circuit.
- Soil Moisture Sensor - The Soil Moisture Sensor will be used to check the condition of the soil.
- Solar Panel - The solar panel is the source of electricity.
- GSM Shield - The GSM Shield is used to establish communication between a user and microcontroller. It requires a SIM (Subscriber Identity Module) card just like mobile phones to activate communication with the network. The purpose of the component to the project is to wirelessly alert the farmers about the condition of the soil.
- Mobile Phones - The farmers will use mobile phone to receive notification from the GSM Shield about the condition of the soil. This helps a lot in determining when to water the plants using the water pump. As mentioned in the study of [12], that GSM Shield Technology is suitable for modern days and improved the old method of collecting data in the farming areas.

3.2. Working Principle
Soil Moisture sensor checks the level of moisture content. The sensor probes are directly planted to the soil and detect the soil moisture value. The sensor value will be the input for the arduino board and process it. After this, Arduino board reads the moisture condition from soil moisture sensor. If it detects normal, wet and too wet conditions the system will not send a message to the farmer. But if it detects dry and too dry conditions then a text message will be sent to the mobile phone of the farmer informing of the soil moisture value, the condition of the soil and the action to be taken. The farmer then will turn

on the water pump to water the plants via SMS. The farmer will send "ON" message to the system to water the plant. Otherwise, "OFF" message to turn off the water pump. Figure 3 shows the block diagram of the developed device for the whole system.

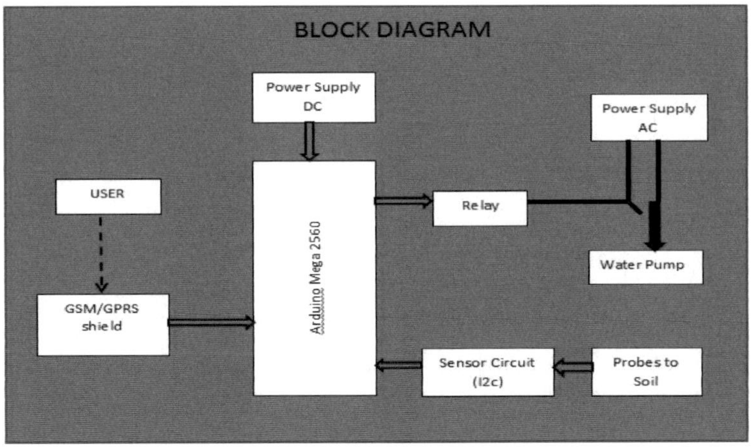

Figure 3. Block Diagram of the developed device

4. Results and Discussion

The Table below shows the values detected during the testing of the equipment in the field at Tagudin, Ilocos Sur, Bangar and Sudipen, La Union. It was validated by five Rice farmers and seven Maize farmers since they are the farmers that are available and doing the watering and checking of the soil condition when the proponents did the testing of the equipment. Too Wet, Wet, Normal, Dry, Too Dry are the characterization used to better understand the soil moisture values.

Based on the gathered data during the testing of the equipment, as shown in the table below from when the Rice was transferred until 80% of growth, it requires plenty of water. But when the Rice plant reaches 81% growth, the farmer's start to reduce the amount of water going to the Rice plants until it will be harvested. Like in the Rice, for example under the Wet, the sensor detected the value 270 to 383 during 0 to 80% growth of Rice and 360-405 during the 81% until it is ready to be harvested. This implies that the value is lesser if the growth stage is 81%, thus, the amount of water needed for Rice is lesser if it reaches 81% growth and until it will be harvested.

On the other hand, Maize plants needs plenty of water from 0 to 80% growth stage compared to when it reaches 81% growth; the required water will start to reduce. Like under Wet, the sensor detected the value 281 to 369 under 0-80% growth, and 351-395 under 81% -100%. This implies that the more water is needed when the growth stage is from 0-80% growth compared to when the Maize crops reaches 81% of the growth stage.

Table 1. Characterization of Soil Moisture Level for Rice and Maize Crops with Fuzzy logic

Growth	Soil Moisture Level Values									
	Too Wet		Wet		Normal		Dry		Too Dry	
Crops	0-80%	81-100%	0-80%	81-100%	0-80%	81-100%	0-80%	81-100%	0-80%	81-100%
Rice										
Max	269	359	383	405	434	418	450	430	1023	1023
Min	0	0	270	360	384	405	435	419	451	431
Maize										
Max	280	350	369	395	440	460	470	461	1023	1023
Min	0	0	281	351	370	396	441	489	471	490

5. Conclusion

The current work aimed to characterize the soil moisture level for Rice and Maize crops using Arduino with fuzzy logic. Based from the obtained result, the proponent concluded that the Soil Moisture Level gathered using the Arduino Mega was accurate based from the validation made by the experts. Also, it was found out that Rice and Maize needs more water if the growth stage is from 0 to 80%. This characterization will help the farmers in determining the required amount of water for Rice and Maize in specific growth stage.

References

[1] Martin E C 2017 *Methods of Determining When to Irrigate, The University of Arizona Cooperative Extension* Available: https://cals.arizona.edu/extension/ornamentalhort/waterquality/irrigation_when.pdf [Accessed: March 2017].

[2] Daryanto S, Wang L, and Jacinthe P 2016 *Global Synthesis of Drought Effects on Maize and Wheat Production* Available: http://dx.doi.org/10.1371/journal.pone.0156362.

[3] Pandey S and Bhandari H 2009 Drought, coping mechanisms and poverty Insights from rainfed rice farming in Asia *Int. Fund for Agricultural Development (IFAD) (2009).*

[4] Oo H G and Udupa N G 2015 Application of soil moisture sensor in mixed farming *Int. Research J. of Eng. and Technol. (IRJET).*

[5] Mubarak S, Khan S, Sahana N, Megha B, and Sujatha S 2015 Automated irrigation system using wireless sensor networks and GSM module *Int. J. of Advance Research In Sci. And Eng. IJARSE* **4(01)**.

[6] Chhinh N and Millington A 2015 Drought monitoring for rice production in Cambodia *Climate Sci. J.* **3(4)** pp 792-811.

[7] Scherer T F, Franzen D, and Cihacek L 2013 *Soil Water and Plant Characteristics Important to Irrigation* (Fargo, North Dakota: North Dakota State University).

[8] Reyes C M, Domingo S N, Mina C D, and Gonzales K G 2009 *Climate Variability, Seasonal Climate Forecast And Corn Farming In Isabela, Philippines: A Farm And Household Level Analysis*.

[9] Gay L, Mills G, and Airasian P 2006 *Educational Research: Competencies for Analysis and Applications* (New Jersey: Pearson Education, Inc).

[10] Richey R C and Klein J D 2005 Developmental research methods: creating knowledge from instructional design and development practice *J. of Computing in Higher Education Spring 2005* **16(2)** pp 23-38.

[11] Lakhankar T, Ghedira H, and Khanbilvardi R 2006 Neural network and fuzzy logic for an improved soil moisture estimation *ASPRS 2006 Annual Conf. Reno* (Nevada Š).

[12] VidyaSagar B 2012 Green house monitoring and automation using GSM *Int. J. of Scientific and Research Publications* **2(5)**.

IOP Conf. Series: Materials Science and Engineering **325** (2018) 012020 doi:10.1088/1757-899X/325/1/012020

Flood Monitoring and Early Warning System Using Ultrasonic Sensor

J G Natividad[1] and J M Mendez[2]

[1]ICT Department, Isabela State University – Ilagan Campus, City of Ilagan, Isabela, Philippines

[2]College of Computer Studies and Engineering, Lorma Colleges, San Fernando City, La Union, Philippines

sirdyowi@gmail.com, janelli.mendez@lorma.edu

Abstract. The purpose of this study is to develop a real-time flood monitoring and early warning system in the northern portion of the province of Isabela, particularly the municipalities near Cagayan River. Ultrasonic sensing techniques have become mature and are widely used in the various fields of engineering and basic science. One of advantage of ultrasonic sensing is its outstanding capability to probe inside objective non-destructively because ultrasound can propagate through any kinds of media including solids, liquids and gases. This study focuses only on the water level detection and early warning system (via website and/or SMS) that alerts concern agencies and individuals for a potential flood event. Furthermore, inquiry system is also included in this study to become more interactive wherein individuals in the community could inquire the actual water level and status of the desired area or location affected by flood thru SMS keyword. The study aims in helping citizens to be prepared and knowledgeable whenever there is a flood. The novelty of this work falls under the utilization of the Arduino, ultrasonic sensors, GSM module, web-monitoring and SMS early warning system in helping stakeholders to mitigate casualties related to flood. The paper envisions helping flood-prone areas which are common in the Philippines particularly to the local communities in the province. Indeed, it is relevant and important as per needs for safety and welfare of the community.

1. Introduction

In most countries in the world, flood had caused damages to properties and it involved a large amount of loss to individuals and governments. During flood, it is important to have efficient flood response operation system to manage all activities among different related agencies.

These last decades, lots of flooding risk technologies has been developed to minimize the danger of flood in inhabited areas. Currently, the Philippine government funded the Project NOAH of the Department of Science and Technology (DOST). They installed Automated Rain Gauges (ARG) and Water Level Monitoring Stations (WLMS) along the country's major river basins (RBs) [1]. However, project NOAH is still under development in which some essential information are not yet available to view in their website.

Most of these technologies being developed commonly apply in weather forecasting, flood detection and monitoring system using sensing devices, modeling software, Internet and mobile technology [2]. However, these systems are usually for one-way communication only. In order to get an update or latest information, local communities need to access the website. And in accessing this website, it requires computer or smart phone that has an Internet feature, and most individual could hardly afford to purchase one. In addition to that, individuals are busy for their daily routine, and monitoring activity cannot be their priority [1]. These are the reasons why communities are blinded with the current status of the nearby river watershed. The unawareness led to the overflow of the watercourses of the river waterway and the subsequent inundation of various localities causing extensive damages to properties and human life.

The City of Ilagan is located at the central portion in the province of Isabela. It is the River Basin of its neighboring towns particularly in the southern portion of the province. Floods caused by these rivers flow down very slowly because of surface retention over the extensive flood plain, extremely gentle slope, retardation of flood by several gorges and river meander [3].

Figure 1. Geographical Map of Flooded Areas in the Northern Protion of Isabela

This paper presents a project that is more localized to help the communities affected by flood in the province of Isabela particularly in the northern area by providing an interactive and real-time information on the current water level in the two majors portion of the province. This project also widens the coverage of people that can receive the information to improve the emergency measures during floods.

Furthermore, this study builds a prototype that detects the current water level across the watershed of Cagayan River and its surrounding areas through ultrasonic sensors. The geographical area was sub-divided into two, where monitoring devices were installed. Specifically, the objectives of this study is to create a flood monitoring system that monitor the water level of the rivers using ultrasonic sensors and to design and develop an early warning system.

2. Architectural Framework

The two monitoring devices are composed of Ultrasonic sensor to measure the distance of the water level, Arduino micro-controller that process the signal from the sensor, GSM module to send the data or information from the micro-controller to the computer server and a power source using Solar Panel, Regulator and Battery. Once a sensor is triggered, an output signal will be relayed to the micro-controller which serves as a switch that triggers the connected GSM module to send an alert message or water level status to another GSM modem connected to a computer server. Then, the developed program installed in the computer server will interpret and analyze the message received then automatically send a text message to the concern agencies' numbers stored in a database. Also, the developed program will then automatically relay the alert message or status by uploading to the developed website. Furthermore, concern agencies, local officials and the local communities could inquire about the current status by sending a message that contains keywords.

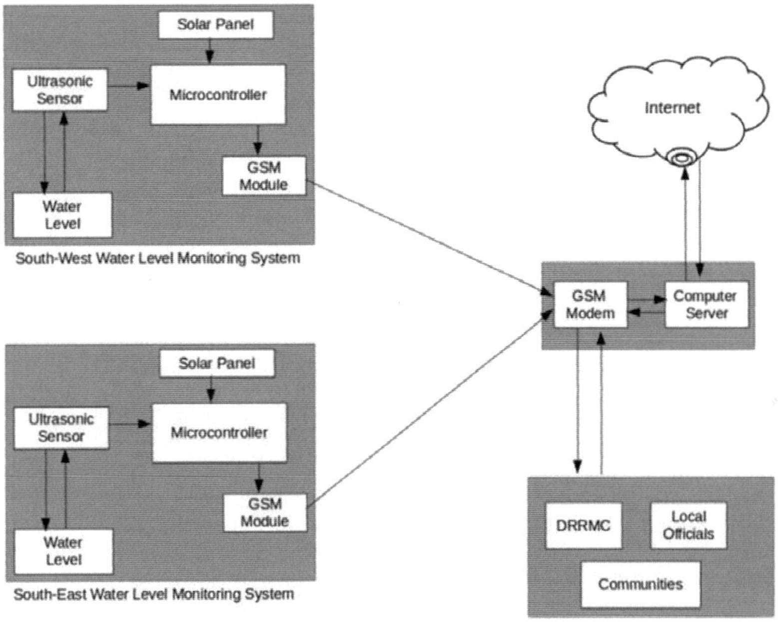

Figure 2. System Architecture

This paper presents the utilization of ultrasonic sensors because of its capability and reliability. Since the Philippines is considered among the most flood prone in the world due to variety of factors, the project NOAH relies on Ultrasonic sensors for water level monitoring. Ultrasonic sensors are deployed on hundreds of coastal tide gauge platforms that provide tsunami and tropical storm surge warning data. They are also deployed on similar platforms that monitor flooding on the different rivers. The newest flood warning system is being deployed to monitor flooding on urban street. And with continues development in ultrasonic sensing, the researchers opted to use this sensor for the project.

The use of GSM also presented in this paper for transmitting data and as mode of communication to the concern stakeholders of this project. Due to its simplicity and availability to the public now-a-days, it is very obvious that information dissemination can be easily achieved. Specifically, the study utilizes

IOP Conf. Series: Materials Science and Engineering **325** (2018) 012020 doi:10.1088/1757-899X/325/1/012020

the use of SMS for the reason that aside for being the cheapest way to avail and transmit information in a remote area, it doesn't require high data bandwidth.

3. Methodology

The ultimate aim is to build a water level detection using ultrasonic sensor to monitor the rivers in the south-east and south-west portion of the province of Isabela and develop a web and SMS application as an early warning system that provides essential information to the local communities and concern agencies.

An SMS approach was used for transmitting data from the monitoring system to the computer server and for sending notification to the concern stakeholders. The SMS application was installed in the computer server to process the received data and make proper action. The application also implement fuzzy logic algorithm for decision making. The inputs of the algorithm are the water level status coming from the two monitoring systems sent through SMS. A threshold value was set in the two monitoring system as basis for the Arduino to trigger the GSM module to send an SMS to the computer server. Then the developed program installed in the computer server send an SMS notification to the concern stakeholders and uploads an update post in the developed web-based monitoring system. After the development of the prototype, the model had undergone several tests and experimentations to check the effectiveness of the system.

4. Result and Discussion

4.1 Flood monitoring system that monitors the water level of the rivers using ultrasonic sensor

The researchers played out a model, test the ease of use and dependability of the developed prototype. It was tried first in a prototype environment that the researchers made and played out the trial. The test decided whether it meets the necessities of the client. The figure below shows the prototype assembled and the connection of the different hardware components.

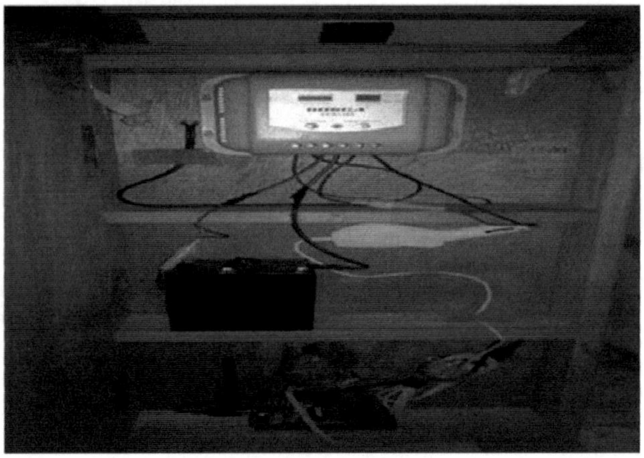

Figure 3. Connection of the different components of the proposed Prototype

The figure shows that in the upper deck is the solar panel that is connected to a regulator. The regulator charges the battery and prevents overcharging to protect against over-voltage. And in the lowest deck lies the Arduino micro-controller topped by the GSM Module and the Ultrasonic sensor facing down detecting the distance of the water.

To be able the developed prototype function properly, a script written in Arduino programming language was uploaded first to the Arduino micro-controller board. The script was tested using the Arduino Integrated Development Environment (IDE) in a temporary environment to check if it meets the expected output.

4.2 Prototype Monitoring System Testing

The researchers tested the developed prototype through a temporary basin to test the level of water. The inputs have several sub-parameters to obtain accurate data. In the designed prototype, water level is measured in inches. The input has four options to consider. When the water level exceed the threshold value set in the script that was uploaded in the Arduino Micro-controller, a notification message containing the level of water and the alert level will be sent to the computer server. The table below shows the different option as input for the monitoring system.

Table 1. Threshold Value Set for Prototype Testing

Water Level (Inches)	Alert Level	SMS Notification Delivery
5" and below	Normal	10 minutes interval
6"-10"	Moderate	5 minutes interval
11"-15"	Critical	1 minute interval
16" and above	Emergency	1 minute interval

4.3 Design and develop an early warning system using Web and SMS.

The researchers developed two different platforms in disseminating information to the concern stakeholders for a possible flood event. One would be the real-time monitoring through a web-based system that can be access through the Internet. Another one is the SMS notification system wherein an automatic communication between the system and the local communities and other concern agencies in the province of Isabela.

The web-based monitoring system was written in PHP programming language and used MySQL as back-end to store information uploaded by the SMS notification system. The web-based monitoring system also contains the different information for monitoring flood such as level of water, alert level, flood warning status, affected areas and update logs. Also the web application automatically updated when new information was uploaded.

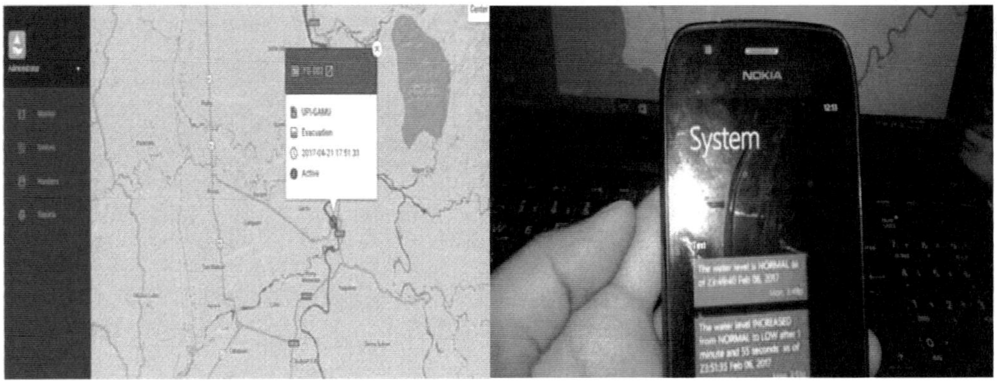

Figure 4. SMS Alert and Web-based Interface

The developed SMS application acts as the brain of the entire system. It performs processing on data sent by the water level monitoring system, responsible in uploading an update for the web-based monitoring, sending notification to the concern stakeholders and reply on request for update through keywords. The SMS application implement fuzzy logic algorithm in giving notifications. Based from the alert level received that served as input, the application can provide a warning status as an output.

Table 2. SMS Application Flood Warning Status

Alert Level	Flood Warning Status	SMS Notification Delivery
Normal	Safe	10 minutes interval
Moderate	-Prepare for Evacuation (Area1, Area2, Area3,..) -Standby (Area6, Area7, Area8,...)	5 minutes interval
Critical	-Evacuate (Area1, Area2, Area3,..) -Prepare for Evacuation (Area6, Area7, Area8,...)	1 minute interval
Emergency	-Evacuate (All Flooded Areas)	1 minute interval

Also, the developed SMS application allows concern individuals to inquire about the current status of a possible flood event. By sending a message containing the keywords "Flood Status", the system will send a reply message that contains the alert level, flood warning status and areas affected by flood. By this feature it allows the system to perform two-way communication between the system and the community in general.

5. Conclusion

The project contributes towards economy and the citizens. It envisions a safe, prepared and less casualty community before, during and after typhoon devastation. The model also promotes the use of real-time monitoring system through the developed web-based application and SMS notification system as an easy medium in disseminating information particularly in the remote areas. By allowing the system in two-way communication, it gives more flexibility in providing important information to the community.

Finally, the developed flood monitoring and early warning system that utilizes ultrasonic sensor to detect water level, functions perfectly according to the specification provided. It successfully passed several tests based on the different parameters.

References

[1] Pagatpat J C, Arellano A C and Gerasta J (Iligan City, Philippines: Mindanao-State University-Iligan Institute of Technology).

[2] Vitales J S, Villajin L D, Destreza F G, Ricafranca D V, and Rodriguez V M R 2016 *College of Engineering and Computing Sciences* (Batangas, Philippines: Batangas State University ARASOF Nasugbu).

[3] dost.gov.ph 2016 Available: http://dost.gov.ph/component/search/?searchword=flood%20monitoring%20with%20project%20NOAH&searchphrase=all&Itemid=101.

[4] Inquirer.com 2016 Available: http://newsinfo.inquirer.net/845239/heavy-rains-flood-bridges-roads-in-isabela-cagayan

[5] Inquirer.net 2012 *Philippines: DOST undertakes flood Prevention and mitigation project* Available: http://goo.gl/i1dd50 [Accessed: January 2012].

[6] Sipalay Flood Control Project 2010 Available: http://goo.gl/UhkDfU [Accessed: December 2010].

Improving Smart Home Concept with the Internet of Things Concept Using RaspberryPi and NodeMCU

Yasirli Amri[1], Mukhammad Andri Setiawan[2]

[1,2] Department of Informatics, Universitas Islam Indonesia, Yogyakarta, Indonesia

[1]13523220@students.uii.ac.id; [2]andri@uii.ac.id

Abstract. The Internet of things (IoT) is getting more tractions in recent years. One of the usage scenario of IoT is smart home. Smart home basically provides home automation for installed devices at home such as thermostat, lighting, air conditioning, etc and allows devices connected to the Internet to be monitored and controlled remotely by user. However many studies on smart home concept focusing only on few main features. They still lack of important usage of IoT i.e. providing energy efficiency, energy monitoring, dealing with security, and managing privacy. This paper proposes a smart home system with RaspberryPi and NodeMCU as the backend that not only serves as home automation and merely a switch replacement, but to also record and report important things to the owner of the house e.g. when someone trespasses the house (security perimeter), or to report the calculation of how much money has been spent in consuming the electrical appliances. We successfully examine our proposed system in a real life working scenario. The communication between user and the system is done using Telegram Bot.

Keywords: Smart Home, IoT, Home security, Home automation

1. Introduction

Housing is one of the primary needs of society. With better economic situation, the demands of having a safe and comfortable home is increasing. Technology is one of the things that able to affect the security and can create comfort of a home. Smart home aims to make homeowners feel more comfortable and safe when he is inside or outside the house. Smart homes once hailed as the domain to the rich but now slowly and steadily becoming a necessity for many people in urban areas due to its ubiquitousness as the price are getting cheaper year by year. The IoT helps smart home adoption in many places, and among the features that attracts many people are security and automatic control features [1].

Smart home is not a new concept. The first smart home technology named X10 was developed by a company in Scotland in 1975 [2]. A smart home appears "*intelligent*" because its computer systems can monitor many aspects of daily living. As the number of controllable appliances in the home rises, the ability of these devices to interconnect and communicate with each other digitally becomes useful and desirable feature [3]. The term Internet of Things (IoT) was first introduced by Kevin Ashton in 1999 [4] where it allows a device that is naturally not connected to the network to be connected to the internet and can be monitored and controlled remotely by user. Most research on smart home is now only focus on one or two main features only, and rarely discusses on the integration of multiple features such as how to achieve energy efficiency, intensive monitoring, security enhancement, and maintaining privacy. To fill the gaps in the IoT research, we design and build a smart home prototype with internet of things concept using RaspberryPi and NodeMCU that integrates enhancement features. We proposed a smart home that is not only serve as a substitute for mechanical switches to turn on or off electrical devices but also record and report important things related to the home owner, such as home security. Our proposed work can make a report if there are people who trespass into the house. The proposed system can also calculate the power that has been used by electronic devices so that homeowners can save money in electricity bills. This greatly helps the homeowner to monitor the house even when he is not at home.

2. Literature Review

Since smart home technology was first introduced in 1975, many researchers have discussed the topic very well. Starting 2010, the smart home topics most widely associated with IoT [5]. Panna et. Al [6] in their research developed the *Development of Energy Saving Smart Home Prototype.* This research discusses the making of a smart home prototype using PIC18F458 microcontroller. The main purpose of this prototype is to save electricity. The prototype only uses temperature and infrared sensors. The temperature sensor is used to regulate the air conditioning and infrared sensors to detect human presence in the room. microcontroller will decide whether the air conditioner and the lights should be on or off based on the sensor readings. The estimated the cost of electricity usage will be recorded into a database that can be accessed by the user.

In other research, Piyare and Song [7] proposed a *Smart Home-Control and Monitoring System Using Smart Phone*. This research discusses the smart home control and monitoring using micro-web server and smartphones. Micro web-server installed on arduino used to transmit data from the sensor to smartphone user, then the user can send commands to arduino to control electronic devices. There are other research discusses the use of microcontroller in smart home environment such as Raspberry Pi Interactive Based Home Automation System through the Internet of Things [8]. This research discusses the home automation system using the Raspberry Pi and MSP430 microcontroller. Raspberry Pi acts as a master and MSP430 as a slave. MSP430 sensor will transmit readings to the Raspberry Pi, and raspberry will decide what commands will be sent to the slave or the other slave.

Other research discusses on how a smart home was built with GSM based automation [9]. This research discusses the home automation system using arduino microcontroller connected to the GSM modem controlled from android smartphone. This automation system is only used to turn on or turn off electronic devices. On another study, smart home was built using email as the platform [10]. This research discusses home security system using facial pattern recognition to get into the house. This system uses a BeagleBone and cameras. The System used email to make communication with users. Another study proposed a prototype of smart home intelligent lighting control architecture onboard sensors using a mobile computing system [11]. This research discusses a prototype of a smart home by using Ambient Light Sensor on smartphones. The readings from these sensors will be sent to the Arduino Uno and forwarded to the Raspberry Pi to decide whether the lights should be on, dim, or off.

From some previous research it can be seen that most smart home systems were made with only one or two functions just like a remote control for an electronic device or an electronic device automation. Therefore, we decided to make a smart home prototype using the Raspberry Pi and NodeMCU. The main features of our prototype made are automation, control, monitoring, and security. The automation will create an automation e.g. lights that is based on human presence. The control feature is able to turn on and off electronic devices remotely. The monitoring feature to determine the approximate cost of electricity from every electronic device. Security to keep the house from thieves when the owner was not at home.

3. Proposed System

The system is developed using the Raspberry Pi and NodeMCU as the main components. The communication between Raspberry with NodeMCU are using MQTT, while the communication between users and Raspberry using a Telegram bot. Other components used in the development of this prototype are sensors that consists of rain sensor, door sensor, passive infrared sensor (PIR), and DHT22 (temperature and humidity) sensor. The proposed block diagram of smart home prototype as presented in Figure 1.

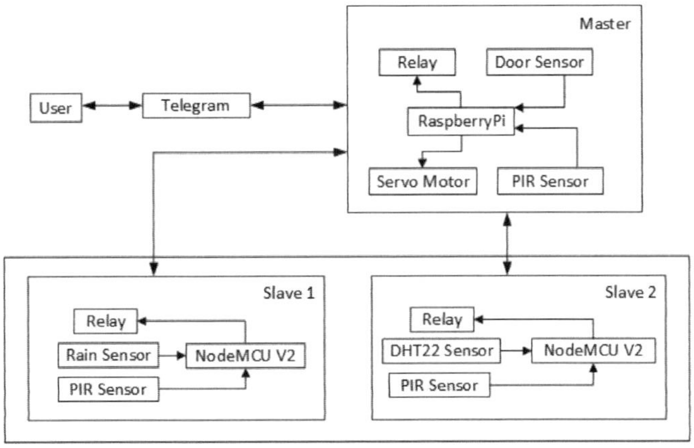

Figure 1. Block diagram of smart home prototype

The system has 3 modes which can be selected by the user; the modes are auto, manual, and secure. Auto mode works by detecting the presence of a human using a PIR sensor and a temperature sensor using DHT22. If the PIR sensor detects human presence in the room, the sensor will transmit the readings to the Raspberry Pi results, then the Raspberry Pi will instruct the relay to turn on the lights that are in the room. DHT22 sensor is used to determine whether the air conditioning should also be turned on or not.

Manual mode works by waiting for user commands given to the system via bot telegram. The secure mode is used when the user is not at home. When secure mode is active, PIR sensor and door sensors will work continuously. The system will send a warning message to the user if any suspicious activity is detected by a sensor such as detecting the presence of human or door is open when no one at home. The locations of sensors and relays on our smart home prototype is presented in Figure 2.

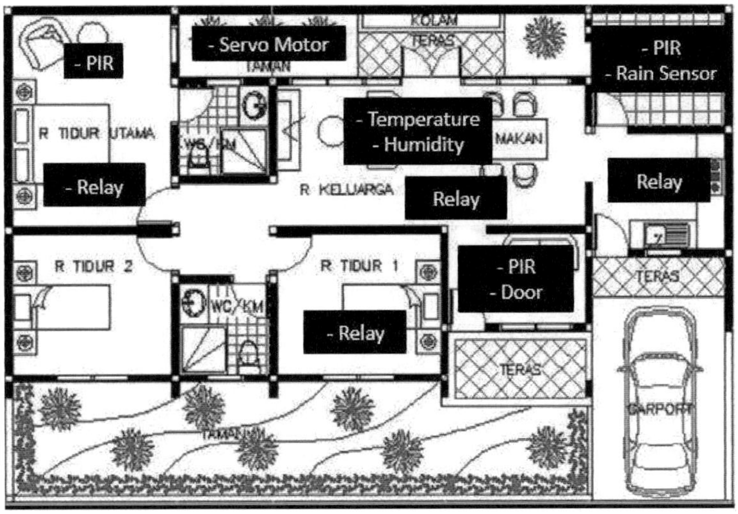

Figure 2. location of sensor and relay on smart home prototype

Because the result of this research is a prototype, it is necessary to do some manipulation for testing purposes. The objective is to get the conditions according to actual events. Conditions and sensor manipulation are presented in Table 1.

Table 1. Condition and Sensor Manipulation

Condition	Sensor manipulation
There are human	Open the PIR sensor cover and Moving the hand in front of the sensor.
There is no human	Installing the PIR sensor cover.
Rainy day	Drops of water on the rain sensor.
Sunny day	Drying the rain sensor.
Turning on and off electronic devices manually	Send a message to telegram bot to turn on or turn off electronic devices.
Temperatures above 30^0C	turn on the smartphone's flash light near the sensor.
Detecting thief	Turn on secure mode and moving the hand in front of PIR sensor.

4. Implementation

We build our prototype with Raspberry Pi as a master and MQTT broker, NodeMCU as a slave, and MQTT as a communication protocol between Raspberry and NodeMCU.

4.1. RaspberryPi 3 Model B

Raspberry Pi is a credit card sized computer. It is a capable little computer which can be used in electronics projects, and for many of the things that desktop PC does [12]. Raspberry Pi used in this research is the Raspberry PI 3 Model B. It is selected because the price is cheap and the amount of community supports. Figure 3 shows Raspberry Pi 3 Model B and its pin diagram.

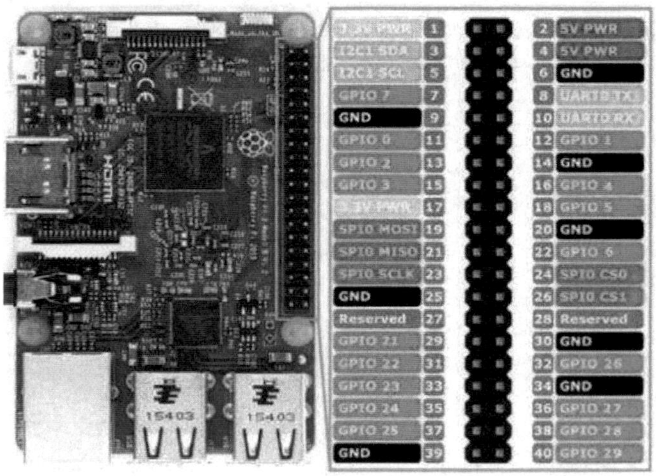

Figure 3. Raspberry Pi 3 Model B

The operating system used on this raspberry is Raspbian OS. After the operating system is installed we then proceed with the installation MQTT so that Raspbery Pi can become MQTT broker. Devices that are installed in Raspberry Pi consists of sensors, relay, and servo motors. Pin Configuration for each device on Raspberry Pi can be seen in Table 2.

IOP Conf. Series: Materials Science and Engineering **325** (2018) 012021 doi:10.1088/1757-899X/325/1/012021

Table 2. Pin Configuration for each device on Raspberry Pi

Device	Pin	Raspberry Pin
PIR sensor	VCC	01 (3.3V)
	GND	14 (GND)
	OUT	04 (GPIO04)
Magnetic Door Sensor	GND	09 (GND)
	OUT	03 (GPIO02)
Relay	VCC	04 (5V)
	GND	06 (GND)
	IN1	18 (GPIO24)
	IN2	16 (GPIO23)
	IN3	12 (GPIO18)
	IN4	10 (GPIO15)
	IN5	08 (GPIO14)
	IN6	-
	IN7	-
	IN8	-
Servo Motor	VCC	02 (5V)
	GND	34 (GND)
	SIGNAL	40 (GPIO21)

After the Raspberry Pi configuration is complete, we then proceed with NodeMCU configuration that serves as slave to transmit sensor readings and receive commands from RaspberyPi.

4.2. NodeMCU V2 ESP8266

NodeMCU first was developed in 2013 by a Chinese company named Espressif makes a System on Chip (SoC) called ESP8266. NodeMCU is IoT platform that is open source and uses the Lua programming language. In its development, NodeMCU can finally be programmed using Arduino IDE so it can run the libraries used by arduino. NodeMCU is selected because of its cheap price and the extent of community support. NodeMCU V2 and its pin diagram can be seen in Figure 4.

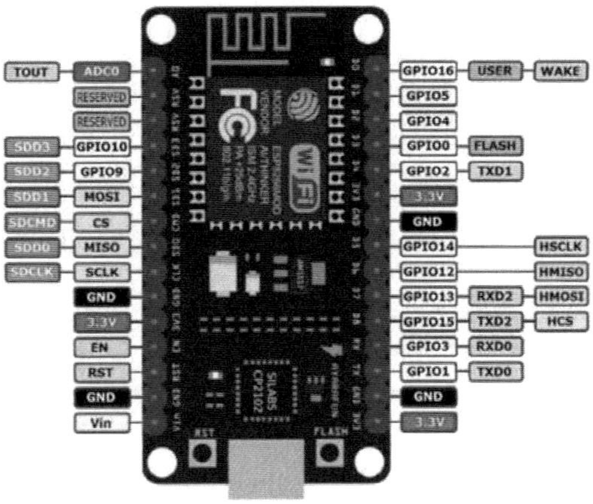

Figure 4. NodeMCU V2 ESP8266

This research uses 2 NodeMCU. First NodeMCU device installed with DHT22 sensors, PIR, and relays. The second one installed with rain sensors, PIR, and relays. NodeMCU pin configurations for each device are presented in Table 3 and Table 4.

Table 3. Pin Configuration for each device on NodeMCU 1

Device	Pin	Pin NodeMCU
PIR sensor	VCC	VCC
	GND	GND
	OUT	D6
DHT22 sensor	VCC	VCC
	GND	GND
	OUT	D7
Relay	SIGNAL	D5
	GND	GND

Table 4. Pin Configuration for each device on NodeMCU 2

Device	Pin	Pin NodeMCU
PIR sensor	VCC	VCC
	GND	GND
	OUT	D8
Rain sensor	VCC	VCC
	GND	GND
	AO	A0
	DO	-
Relay	VCC	VCC
	GND	GND
	IN1	-
	IN2	D2
	IN3	D3
	IN4	D4
	IN5	D5
	IN6	D6
	IN7	-
	IN8	-

After NodeMCU configuration is complete. Then proceed with installing MQTT in Raspberry Pi so it can communicate wirelessly with NodeMCU.

4.3. Message Queue Telemetry Transport (MQTT)

Message Queue Telemetry Transport (MQTT) is a Machine-to-Machine (M2M) / "Internet of Things" connectivity protocol. MQTT is selected because it is a lightweight messaging protocol. MQTT send and receive messages in a very small size. MQTT use publish and subscribe concept to send and receive message, it is client and server based. Server which is better known as the broker receives a message from a client, it can be data message publish or subscribe. MQTT is not a new protocol, created by Dr. Andy Stanford-Clark from IBM and Arlen Nipper from Arcom (now Eurotech) in 1999[13]. MQTT publish-subscribe architecture can be seen in Figure 5.

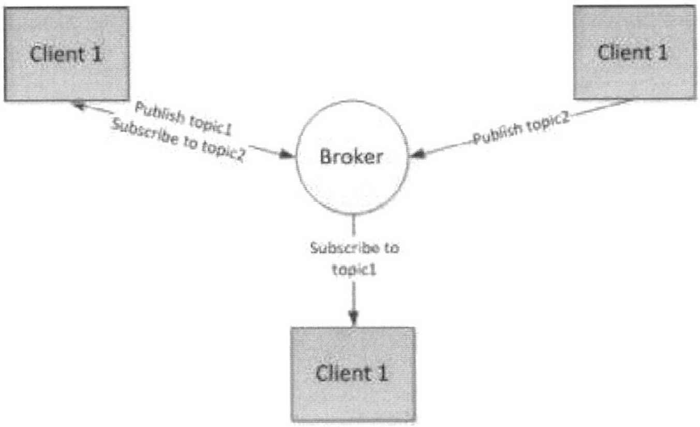

Figure 5. MQTT publish and subscribe architecture

TO increase the reliability, MQTT offered three types of modes which are called Quality of Service (QoS) [14], Detailed information about each QoS MQTT can be seen in Table 5.

Table 5. Various QoS on MQTT

QoS	Information
0	Messages are sent once and does not require confirmation or Acknowledge (ACK).
1	Ensuring that messages are sent at least one time and requires ACK.
2	Ensuring that messages sent exactly once.

Once the configuration is complete and Raspberry can communicate with NodeMCU via MQTT. The next step is to install the entire prototype model. Figure 6 shows our final prototype that already assembled.

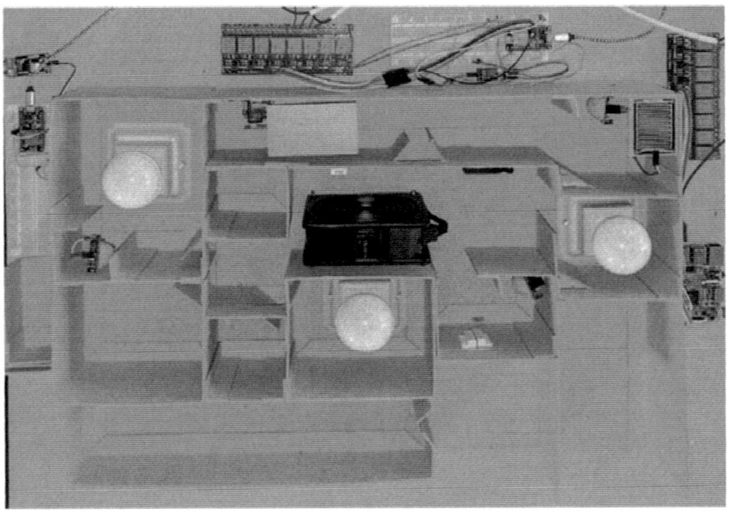

Figure 6. Prototype Smart Home

5. System Testing and Analysis

After the configuration and the assembly of each device on a prototype smart home is completed, we tested with multiple scenarios based on actual events. The conditions and the test results can be seen in Table 6.

Table 6. The conditions and results of testing

Condition	Supposedly output	Test result
User are outside. User select auto mode. User enters the living room. The temperature in the living room 26^{0}C.	Living room lights automatically on.	valid
User are in the bedroom. User selects auto mode. There is a clothesline in the backyard. Users walk into the living room and backyard Rainy day.	Lamp in living room and backyard automatically on. Roof closed.	valid
User are in the living room. User selects auto mode. The weather was hot and the room temperature of 30^{0}C.	Lights and AC automatically on.	valid
User are in the living room. User change auto mode to manual mode. Due to an error input, the user must change the power and lifespan e6 (lights) on the database to 12 Watt power with lifespan 15,000 hours, from the previous 8 Watt power with lifespan 13,000 hour. User turn off and then turn on the lights.	Light off then back on, and the calculation of power and lifespan of the lamp is using the latest input.	valid
User are in the bedroom. User manually turn on the lights. Users get out and turn off the lights manually.	The light was on and then off. Once the lights are off, the user will receive a message from the system that said the lights are off with the remaining lifespan of the lamp.	valid
User are in the bedroom. User selects secure mode. User running toward the backyard.	Users detected as a thief.	valid

We then analyse the usability of our system and classify both advantages and disadvantages of the system as presented by Table 7.

Table 7. Advantages and disadvantages of the system

Advantages	Disadvantages
• Replaces the manual switch at home. • Noting the estimated cost of electricity for every electronic device. • Improve home security. • The system does not require a great power. • Users can control and monitoring the house remotely.	• Command to the system by sending a message, that is not practical for the user. • Telegram bot for the system can be used by many users, but should be included in the same group chat so that everyone can send commands to the system.

6. Conclusion

Based on our testing and analysis, it can be concluded that our prototype works well. The main function of this system is automation, control, monitoring, and security. Automation and remote control enables users to operate the electronic device even when users are not at home. Monitoring function can help users to estimate the cost of electricity usage for any electronic device. Security functions may provide a sense of security to users when they're not at home because the system will work continuously and report any suspicious activity that detected by the sensor. It is expected that this prototype can be the basis for further smart home development with better security and more functions.

References

[1] Vyas C 2016 Smart home analysis in india : an iot perspective **144(6)** pp 29–34.
[2] Robles R J and Kim T 2010 A Review on security in smart home development *Int. J. Adv. Sci. Technol.* **15** pp 13–22.
[3] Hamed B 2012 Design & implementation of smart house control using LabVIEW *Int. J. Soft Comput. Eng.* **1(6)** pp 98–106.
[4] Ashton K 2010 That 'Internet of Things' thing *RFID J.* p 4986.
[5] Alaa M, Zaidan A A, Zaidan B B, Talal M, and Kiah M L M 2017 A Review of smart home applications based on internet of things *J. Netw. Comput. Appl.*
[6] Panna R, Thesrumluk R, and Chantrapornchai C 2013 Development of Energy saving smart home prototype *Int. J. Smart Home* **7(1)** pp 47–66.
[7] Piyare R and Seong R L 2013 Smart Home-control and monitoring system using smart phone *Proc. 1st Int. Conf. Converg. it's Appl.* **24** pp 83–6.
[8] Ganesh S, Venkatash, Vidhiyasagar, and Maragatharaj 2014 Raspberry Pi based interactive home automation system through Internet of Things *Int. J. Res. Appl. Sci. Eng. Technol.* **3** no. November 2015 pp 277–80.
[9] Jivani M N, "GSM Based Home Automation System Using App-Inventor for Android Mobile Phone," *Int. J. Adv. Res. Electr. Electron. Instrum. Eng.*, vol. 3, no. 9, pp. 12121–12128, 2014.
[10] Rajiv P, Raj R, and Chandra M 2016 Email based remote access and surveillance system for smart home infrastructure *Perspect. Sci.* **8** pp 459–61.
[11] Tang S, Kalavally V, Yew K, and Parkkinen J 2017 Development of a prototype smart home intelligent lighting control architecture using sensors onboard a mobile computing system *Energy Build.* **138** pp 368–76.
[12] RaspberryPI 2017 Raspberry *Pi FAQs - Frequently Asked Questions* [Online] Available: https://www.raspberrypi.org/help/faqs/. [Accessed: 05-Apr-2017].
[13] Prada M A, Reguera P, Miguel A, Mor A, Fuertes J J, and Dom M 2016 *Communication with resource-constrained devices through MQTT for control education* pp 150–5.
[14] Amaran M H, Arif N, Noh M, and Rohmad M S 2015 A Comparison of lightweight

Communication protocols in robotic applications *Procedia - Procedia Comput. Sci.* **76** no. Iris, pp 400–5.

IOP Conf. Series: Materials Science and Engineering **325** (2018) 012022 doi:10.1088/1757-899X/325/1/012022

Automatic Speed Control and Turning ON/OFF for Smart Fan by Temperature and Ultrasonic Sensor

Keeratiburt Kanchanasatian

Department of Computer Science, Faculty of Science and Technology, Rajamangala University of Technology Thunyaburi

keeratiburt_k@rmutt.ac.th

Abstract. When it is hot, using a fan is an inexpensive choice compared to spending more on using an air conditioner especially in some areas where high temperature is naturally normal. Some problems, nonetheless, arisen such as users forget to turn off when being away, some people might get a feeling of annoyance and perhaps frustration when getting up from the seat in order to adjust the speed. These result in a rise of electricity bill from leaving the fan on when it is not in use and it may cause disasters, fire for example, when the motor get heated up. A prototype of smart fan was built in this research using ESP8266 as a microcontroller, DHT22 and HC-SR04 are used to measure temperature for speed control and detect the user for automatic on/off respectively. A group of participants used the prototype and they were interviewed to give some feedbacks, comments, and suggestions from the experiences after using it. The results show that they were satisfied from the automation; it gave the sense of staying in a modern house with an automatic wind blower. It also assisted to reduce energy consumption according to target group. Some users stated that the environment and atmosphere in the room was not too hot because the fan was working in the background. This has shown that the prototype significantly gave the participants a feeling of relaxation and comfortability and also was a part of energy and cost reduction.

1. Introduction

Fans are commonly used in places where the temperature is normally. Also the concept of Smart Home are getting popular all around the world even though the concept was known for decades [1].The problems arise when all the fan's speed are adjustable; however, this can be done manually. It means that if the user wants to change the speed, they have to move to the fan the push the button by themselves. Another drawback is that many people forget to turn the fan off after they have finished using it. They just leave the place and do not notice that the fan is left opened. This has resulted in the increase of the electricity bill and 5-10% of electricity consumed is wasted [4]. Sometimes it may cause a fire because the fan works by the rotation of a motor; if the motor keep operates for a long time, it produces the heat and the fan body is made of plastic which can be set fire quite easy. A system that is controlled by remote device such as remote control is a better idea due to the ease of non-contact to the system directly [6], [7].

The system that will be developed in this research will be called "Smart Fan". The smart fan can adjust its speed level according to the room temperature. Additionally, it starts working when the user sits in place such as on a sofa or chairs. After the user leaves the seat, the fan should stop working. This was supposed to bring the user comfortability and energy reduction consumption.

The phrase "Smart Home" has become more popular and common to the public after the research in software and hardware for decades. Technologies used nowadays provide human easier for do the living than before. Homes that are smart are expected to provide the owner automation such as automatic lighting and security system [17], [19].

2. Review

2.1. Hardware

To allow any home appliances to be intelligent, it needs a controller as its brain [4]. Microcontrollers are the control unit needed in this system and Arduino is by far the best choice [5] for development the prototype and also the price is reasonable. A smart appliance not only can make the user feel comfortable but also can assist the elderly or disabled [3]. The problem of wasted energy results in a quarter of all the causes of the greenhouse effect [6]. Some research in DHT22 and HC-SR04 were reviewed to gain knowledge about the two. The DHT22 seems to be a suitable for temperature monitoring as it produces digital output and the price is reasonable [9].

As for the motion sensor, first, passive-infrared sensor (PIR) which emit infrared signal and detect any movement that cuts through its signal. Second, ultrasonic sensor sends out a sound wave and once the wave contacts an object such as human or wall, the wave bounces back to the sensor and the time which the wave has traveled forth and back is calculated to be the distance between the sensor and the object. The difference is that if the object stays still within the infrared signal for a period of time, the sensor will understand that there is nothing within the range because there is no movement anymore [10], [11], [12], [13]. That is, PIR sensor can only detect movement when human come to the front but if the human sits still or falls asleep the PIR sensor will not detect any movement anymore and the fan is turned off. Ultrasonic sensor, in contrast, will always produce the output that there is still an object within a range because the distance is measured all the time. This allows the ultrasonic sensor to detect whether there is an object within its range or not even the object is not moving.

2.2. Literature Review

Ultrasonic sensor was more suitable than PIR sensor in the sense of human detection for the scope of this research because the sensor detect human in one direction which is in front of it but the PIR sensor detection range is in cone shape [13]. Another smart fan from T. Ishrat et al (2014) did not use a microcontroller that allowed the system to connect to the Internet for the concept of the Internet of Thing (IoT) but the prototype in this research used NodeMCU which is able to connect to the network allowing the system to be monitored for managed online for the future approach [13].

Another research from [14] used infrared signal to detect human in a museum. The microcontroller PIC16F887A [14] which could not connect to the computer network compared to NodeMCU used in this research. The fan from [14] was not described whether the fan continued to operate if a person is standing still or not. However, the prototype in this research is guaranteed to be working as long as a person is still there even when the person is sleeping.

As for the power usage monitoring, smart appliances that have a feature of automatic ON/OFF system will save electricity and time [19]. Smart fan from other research were concluded to help on energy conservation and cost reduction [13], [14].

The outstanding difference between the result in this research and the others is that there were no feedbacks, comments, and suggestions from the user. This paper provide those information gathered from participants by the method of interview which is a valuable information to make future process. From the review, the prototype in this paper and the result had more potential in many aspect such as human existence detection, cost reduction and power conservation, and user feedbacks.

3. Methodology

Two main tasks were designed and created to obtain the result in this research, which were interview forms and the system:

3.1. Interview

The criteria of the interview consists of five main parts, which are features, usefulness, comfortability, satisfaction, and feedback. There were five participants testing the system by given the fans developed in this research. The period of using the fans by the experimental group was approximately six months. The data was collected by three interviews. The first interview was two weeks after getting the fans and the questions were about the stability and errors of the fans. The reason these questions were used because the system must perfectly function without any errors otherwise the result of the research will not be concise. This interview took place at the participants' house. The second interview was about the satisfaction from using the fans, are there any features that should or should not be added to the system. The last interview was also about the satisfaction of the user from using the fans. They were asked whether they wanted to keep using it or not.

3.2. The System Prototype

The system consists of three main logical hardware parts, the fans, the sensors and the controllers. All the parts were connected to one another by electric wires. The algorithm was designed and programed in C++. The diagram of the system can be seen in Figure 1.

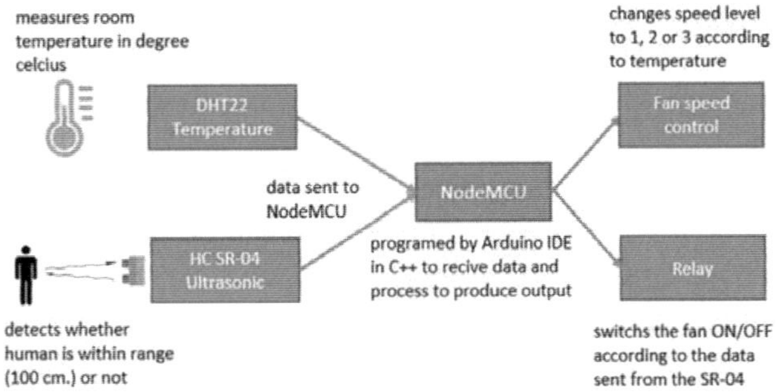

Figure 1. The System Diagram

3.2.1. Fan

Two kind of fans were used in the experiment. The first was the table fan which was placed on any table. And the second was taller and could only be placed on the floor. Both models had three speed levels consisting of four buttons, ON/OFF, speed 1 (slowest), speed 2 (medium), and speed 3 (fastest).

3.2.2. Microcontroller

ESP8266 NodeMCU was used as the controller unit. There were, at first, actually 2 microcontrollers to choose from, Arduino UNO and ESP8266. The differences between the two are the UNO does not have a built-in Wi-Fi chipset [2] and the size is about two times larger than the other one. The NodeMCU was chosen because the size mattered; it must fit inside the fans' body and the smaller circuit blue print allows the entire system to consume less power. Although the Wi-Fi was not in the requirement in this research however for the future work it is already planned to develop a step further to allow the connection of the Internet. The only disadvantage of the NodeMCU compared to the UNO is that the NodeMCU uses a micro USB as a connection port in order to connect to the computer to be programed and code embedded; the UNO, on the other hand, uses the much more stable connection which is type-b USB connection port.

The microcontroller digitally received the temperature data in degree Celsius (^{o}c) from the temperature sensor (DHT22). The level of the temperature was set to three levels. If the temperature was less than or equal to 32°c, the speed was set to level 1 which is the slowest; If the temperature was less than or equal to 35°c, the blade's speed was set to level 2; and if the temperature was more than 35°c, the speed was set to highest level which is the fastest of the fan blade. That is, the fan blade spins faster as it getting hotter automatically. The natural environment of the participants' house was between 30 to 38°c.

3.2.3. Sensors
DHT22 temperature sensors were used to monitor the temperature. HC-SR04 ultrasonic sensors were used to detect the users. The sensors faced at the same direction as the fan. The ultrasonic sensors emits the sound wave and it bounces back to calculate the time the wave travels and it results in the distance between the sensors and the object. The distance was set at one hundred centimeters. That is, if there is a person comes into the wave length, the sensor sends the data to the controller and the controller signals the relay to connect the AC current for the fan allowing it to start working. If the person is at the distance of less than one hundred centimeters, the fan is still working. Once the person is away from the length which more than one hundred centimeters; the fan stops working after the delay of twenty seconds.

3.2.4. Relay
The fan uses 220 AC voltage but the NodeMCU and all the sensor uses 3.3-5 DC voltage which cannot be wired up together. The relay acted as a switch for the fan when the ultrasonic sensor detected the human which was the only one task for it. A 4-channel relay was used because 3 channels were wired to connect to the 3 fan speed level switches and 1 channel was used for turning the fan ON/OFF.

3.2.5. Software
Arduino IDE was used to program all the codes and the programming language was C++.

4. Result
The result was that 100% of the participants were satisfied with the prototype. The scale were from 0 to 5 which 0 means not satisfied at all and five means most satisfied. The prototype were useful according to the result as all of the target group gave number 5 for the usefulness. As for the comfortability, 75% chose it was comfortable but 25% said that it could be difficult to say that using the prototype is comfortable or not because she did not control the fan at all as the fan controlled itself on turning on and off. This seem to be comfortable because if the user feels normal or does not feel annoyed and frustrated from the fan, it means that the fan works in a background and gives the user and feeling or comfort. The features and feedbacks will be described in the next paragraph.

The opinions from all the participants are significantly useful such as the fans should automatically stop working when the temperature drops below 26 because it means that the air conditioner is on. Two participants said that it was quite dangerous to put the plug in all the time because the fan turned on and off by itself and this made the users to forget about the plug. Keeping the plug into the wall all the time can cause a fire if the wire or the plug is in a poor condition. The feeling of the cool wind-blown automatically from the fan after sat down on the chair gave all the participants a new experience of using fans. Furthermore, the fans stopped itself when the users are away were very impressive. One participant recommended that the fan should face the users all the time; it means that some kind of user tracking system must be included into the system to allow this to work. However, this was not the scope of the research but it was a valuable opinion. Another recommendation was that there should be a portable version of the fan with battery for outdoor activities. All the participants emphasized that the system should allow the user to change from auto mode to normal mode to control the fan manually. Even though the fan is basically controlled by hand but the function was taken out because it was designed to be fully automatic. The fans are also suitable for any houses that have elderly because it requires less time to operate the fans.

The NodeMCU itself broke down very easily as twenty of them were used to build up to ten fans. This was because the connection and the program uploading issue. Another reason was that the USB port and cable were plug in and out many times during the testing process. Despite all the problems, the NodeMCU was suitable and perfectly worked because of its size and low energy consumption.

5. Conclusion

The system built in this research as a prototype is an ideal fan for many people. As they live in the world that such a system does not exist, they may not have a thought to want one; nonetheless, once they have tried it, they are satisfied and appreciated by its advantages and comfortability. Even though the hardware that was used in this research is not for enterprise level and not good enough for the market, but there are better ones and the features of the fan attracted the participants to continue using it.

Although this system does not have a database because it is just a home appliance. However, it can be put together for future work to collect data usage and analysis for any purposes such as users' behavior, energy consumption prediction and historical data usage in specific periods of time for energy consumption reduction. An alternative is using Zigbee [10] as Smart Home microcontroller because it consumes even less energy [4] than NodeMCU and can lasts for months. Also smart phones can be included in the system to virtualize data by graphs or notifications to inform the user about their usage because smart phones are things that carried with the people at almost all the time [16].

Lastly, what is learned from all the feedbacks was that any systems should be built based on the consumers' needs, preferences, and favorites in order to make the best out of it.

Acknowledgement

This research is based on the knowledge of all the referencing paper and websites. In addition, the author sincerely thanks all the participants who nicely cooperated and were eager to use the prototype and gave the valuable feedback. The author also thank the department of Computer Science, RMUTT, which provided the facilities for implementing the research.

References

[1] Jiang L, Liu D, and Yang B 2004 Smart home research *Proc. of 2004 Int. Conf. on Machine Learning and Cybernetics* (IEEE Cat. No.04EX826) **2** pp 659-63.

[2] Shkurti L, Bajrami X, Canhasi E, Limani B, Krrabaj S, and Hulaj A "Development of ambient environmental monitoring system through wireless sensor network (WSN) using NodeMCU and "WSN monitoring"," 2017 6th Mediterranean Conference on Embedded Computing (MECO), Bar, 2017, pp. 1-5.

[3] Basanta H, Huang Y P and Lee T T 2017 Using voice and gesture to control living space for the elderly people *2017 Int. Conf. on Syst. Sci. and Eng. (ICSSE), Ho Chi Minh City, Vietnam* pp 20-3.

[4] Kaiwen C, Kumar A, Xavier N and Panda S K 2016 An intelligent home appliance control-based on WSN for smart buildings *2016 IEEE Int. Conf. on Sustainable Energy Technologies (ICSET), Hanoi* pp 282-7.

[5] Alam M R, Reaz M B I and Ali M A M 2012 A Review of smart homes—past, present, and future *IEEE Transactions on Systems, Man, and Cybernetics, Part C (Applications and Reviews)* **42(6)** pp 1190-203.

[6] Golzar M G and Tajozzakerin H 2010 A new intelligent remote control system for homeautomation and reduce energy consumption *2010 Fourth Asia Int. Conf. on Mathematical/Analytical Modelling and Computer Simulation, Kota Kinabalu, Malaysia,* pp 174-80.

[7] Bos B, Chmielewski L, Hoepman J H, and Nguyen T S 2007 Remote Management and secure application development for pervasive home systems using JASON *Third International Workshop on Security, Privacy and Trust in Pervasive and Ubiquitous Computing (SecPerU 2007), Istanbul* pp 7-12.

[8] Zhong D, Ji W, Liu Y, Han J, and Li S 2011 An improved routing algorithm of Zigbee wireless sensor network for smart home system *The 5th Int. Conf. on Automation, Robotics and Applications, Wellington* pp 346-50.

[9] Syafii, Rusydi M I, Putra R, and Putra M H 2016 Real-time measurement of grid connected solar panels based on wireless sensors network *2016 Int. Conf. on Sustainable Energy Engineering*

and Application (ICSEEA), Jakarta pp 95-9.
[10] Bai Y W, Cheng C C and Xie Z L 2013 Use of a time-variation ultrasonic signal and PIR sensors to enhance the sensing reliability of an embedded surveillance system *2013 26th IEEE Canadian Conf. on Electrical and Computer Eng. (CCECE), Regina, SK* pp 1-6.
[11] Bai Y W, Cheng C C and Xie Z L 2013 Use of ultrasonic signal coding and PIR sensors to enhance the sensing reliability of an embedded surveillance system *2013 IEEE Int. Syst. Conf. (SysCon), Orlando, FL* pp 287-91.
[12] Bai Y W, Li Z H, and Xie Z L 2010 Enhancement of the complement of an embedded surveillance system with PIR sensors and ultrasonic sensors *IEEE Int. Symp. on Consumer Electronics (ISCE 2010), Braunschweig* pp 1-6.
[13] Ishrat T, Rahaman M A, and Ahammad A 2014 Smart fan for human tracking *2014 9th Int. Forum on Strategic Technol. (IFOST), Cox's Bazar* pp 124-7.
[14] Mishra R, Raza S, Zulquarnain, Arya R, and Kumar P 2013 Development of automatic person detection system to control ac fan and room lights *Int. J. of Innovative Research in Sci., Eng. and Technol.* **2** Issue 3 ISSN:2319-8753.
[15] Tsai W C, Shih Y J, and Tsai T M 2016 IoT-type Electric fan: remote-controlled by smart-phone *2016 Third Int. Conf. on Computing Measurement Control and Sensor Network (CMCSN), Matsue* pp 12-5.
[16] Tsai W C, Shih Y J, and Tsai T M 2016 IoT-type electric fan: remote-controlled by smart-phone *2016 Third Int. Conf. on Computing Measurement Control and Sensor Network (CMCSN), Matsue* pp 12-5.
[17] Malche T and Maheshwary P 2017 Internet of Things (IoT) for building smart home system *2017 Int. Conf. on I-SMAC (IoT in Social, Mobile, Analytics and Cloud) (I-SMAC), Palladam, Tamilnadu, India* pp 65-70.
[18] Latif S S and Osborn W 2017 Database system design for energy economization of smart home energy services *2017 IEEE Int. Conf. on Smart Energy Grid Eng. (SEGE), Oshawa, ON, Canada* pp 281-5.
[19] Malche T and Maheshwary P 2017 Internet of Things (IoT) for building smart home system *2017 Int. Conf. on I-SMAC (IoT in Social, Mobile, Analytics and Cloud) (I-SMAC), Palladam, Tamilnadu, India* pp 65-70.

IOP Conf. Series: Materials Science and Engineering **325** (2018) 012023 doi:10.1088/1757-899X/325/1/012023

Developing and Validating the Socio-Technical Model in Ontology Engineering

Mesnan Silalahi[ab1], Dana Indra Sensuse[a2], Yudho Giri Sucahyo[a3], Izzah Fadhilah Akmaliah[a4], Puji Rahayu[a6], Elin Cahyaningsih[a7]

[a]e-Government and e-Business Research Lab, Faculty of Computer Science, University of Indonesia

[b]Research Center for S&T Development, Indonesian Institute of Science, Jakarta, Indonesia

Abstract. This paper describes results from an attempt to develop a model in ontology engineering methodology and a way to validate the model. The approach to methodology in ontology engineering is from the point view of socio-technical system theory. Qualitative research synthesis is used to build the model using meta-ethnography. In order to ensure the objectivity of the measurement, inter-rater reliability method was applied using a multi-rater Fleiss Kappa. The results show the accordance of the research output with the diamond model in the socio-technical system theory by evidence of the interdependency of the four socio-technical variables namely people, technology, structure and task.

Keywords: Socio-Techincal, Ontology Engineering, Model, Validation, Meta-Ethnography

1. Introduction

Building ontology is initially regarded as a craft rather than an engineering activity [17]. Some of the methodologies developed in the engineering of ontology were initially developed based on the experiences in building ontology whereby mostly based on the principle of common senses such as OTK methodology [19]. The experiences were organized and structured systematically to become a guideline for ontology engineering (OE). Several other methodologies are developed based on the theory that exist in a particular context or an approach with specific orientation, such as HCOME [18], UPON [15], Onto-Eng.[1], and DILIGENT [16]. The developed methodology partially focuses on either technical approach or social and human approach. There is little known about methodology which are technically focused while still considering the social aspect equally in the approach. This study bridges this gap by building a methodology that equally treat the technical and social aspect in ontology engineering, based on a systematic review of the literature in ontology engineering and an empirical investigation in case studies in the development of an ontology. The review includes analyses on methodologies in building ontology, the tools, the environments and the organization, and the study of principles of design from the perspective of socio-technical system. Based on experiences in building ontology a systematic design method is compiled and proposed in ontology engineering methodology with socio-technical approach that provides a comprehensive guideline to building an ontology. This study has two research questions that is “how to build a methodology in OE based on social-technical approach and how to validate the OE model”.

2. The Socio-Technical Approach in Ontology Engineering

Ref. [7] formulated four interdependent variables in the diamond model of socio-technical system namely: task, people, technology, and structure as shown in Fig. 1 by the arrows, so that changes to one variable will affect other variables.

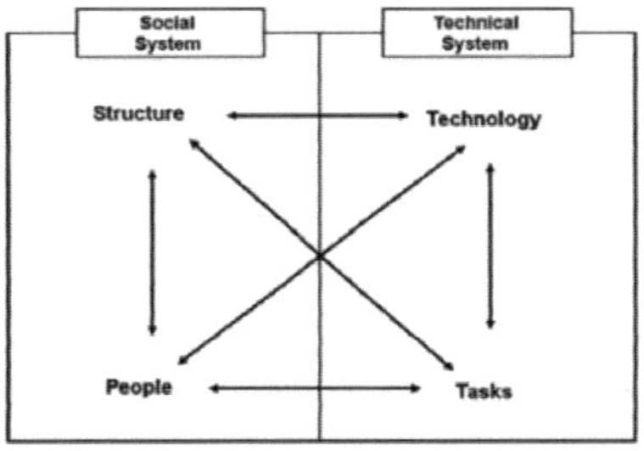

Figure 1. A diamond model in a socio-technical system

Technology is regarded as a tool that assists organizations in completing the work through the mechanisms of transformation from inputs into outputs. The socio-technical ontology engineering perspective would require all activities that support social subsystems (a human capital with its competencies and preferences in a social process) as well as the technical subsystem (engineering functions such as technology and task). A socio-technical approach in ontology engineering has been proposed partially in several other methodologies. The ways of approaching the ontology engineering socio-technically are such as involving domain experts and/or end-users as the central actor, such as in DILIGENT [14], HCOME [6], and HOLSAPPLE [4]. These methodologies proposes to begin the development process not from scratch (using a seed model), as well as proposing to build an ontology by reusing existing ontologies. The context of decision-making in collaborative work is found in [5] which proposed a process through a consensus mechanism. Ref. [9] uses a strategy in building a consensus in ontology engineering in the context of collaborative networks. Ref [13] gives a detailed description of how to build this consensus in NEON methodology as part of the specification and conceptualization process. However, their approach is more a technological support rather than a social approach as it focused on using wiki technology in the negotiation on a higher level. In our study a socio-technical approach to ontology engineering (STOE) methodology is developed. The social aspect is considered equally with the technical aspects in all phases of the ontology development starting from a planning phase until the evaluation phase. Some of the advantages gained by this approach is the knowledge sharing that can be encouraged as early as possible in the development phases, a best practice proposed in [4].

3. Methodology

In our study the initial methodological model was first proposed based on a comparative analysis and by conducting a critical review on several relevant methodologies in ontology engineering. Next the proposed model is triangulated through a meta-ethnography study. The initial model development of the methodology was performed by synthesizing qualitative research data from literatures as identified, searched and selected. The method used in the synthesis of qualitative research publication data in ontology engineering is meta-ethnography. Meta-ethnography as proposed by [8] has seven steps as follow [2]: (1) getting started, (2) deciding what is relevant to the initial interest, defining the focus of the synthesis; locating relevant studies; making decisions on inclusion; and quality assessment, (3) reading the studies, (4) determining how the studies are related, (5) translating the studies into one another, (6) synthesizing translations, and (7) expressing the synthesis.

We investigate the selected literature by doing translation, comparing, analysing and synthesizing. Inductively identification of the key concepts and phrases of words is carried out through thinking deeply and rereading of the text carefully. The emergence frequency of translated words to concept in the dataset is a proxy to the conceptual model. The result from each translation and the synthesis is validated internally using Kappa method through member check which is aimed to avoid subjectivity in the study. The inter-rater reliability (IRR) is used as a measure for the validity of the interpretation. To understand whether observers should have the same opinion, Kappa coefficient K is used as inter-observer reliability estimation which is defined as:

$$K = \frac{Pa - Pe}{1 - Pe} \quad (1)$$

whereby P_a denotes the observed percentage of agreement, and P_e denotes the probability of expected agreement due to chance. The factor ($1\text{-}P_e$) gives the degree of agreement that is achievable above chance, and ($P_a\text{-}P_e$) gives the degree of agreement actually achieved above chance. If the raters are in complete agreement then $K = 1$. Testing using Kappa method is carried out carefully with respect to prevalence and bias that may occur from the observations. IRR were obtained using three observers, which is the minimum number allowed in Fleiss Kappa method (multi-rater).

4. Results

Deductively and structurally, concepts (categories) were acquired in socio-technical system based on the model proposed by [7], which consists of four variables: people, technology, tasks and structure. Using these four variables then we do the synthesis (inductive) from the literature to get the concepts or themes (metaphors). Table 1 shows the results from the synthesis process.

Table 1. An excerp from the results of the synthesis processes

Findings / Evidences	Themes / Metaphors	Concept	Category	Direct Affected Category	Phase
"The users emphasized the need for local control over their ontologies"	Local control	Work procedures	Structure	People	Analysis, Design
"They asked explicitly for a system without a central server, where knowledge sharing was integrated into the normal work, but where different kinds of information, like files, emails, bookmarks and addresses could be shared with others"	Collaboration integrated into normal work	Distributed Collaborative Work	Structure	Technology	Planning
"they commented, that they would share more files at a later stage, when they would feel more confident with the system"	Need confidence in sharing information	Self-efficacy	People	Task	Analysis, Design
"The new users behaved in a similar way as the users in the first stage and did not share many folders, as they wanted to gain confidence in the system first"	New user first of all, gain confidences	Self-efficacy	People	Task	Analysis, Design

The findings showed that none of the socio-technical variables (concepts) show up directly in the frequency count, but most show up indirectly which we called themes or metaphors (e.g., support tools). However, when we explore the keyword frequency using the concepts from our classification model we find that all of the socio-technical variables in our model are represented in the literatures to varying degrees, with the technology variable being the most represented, followed by people then structure. From the results of the synthesis, a cross tabulation is carried out into category along with its themes of the phases identified. The cross-tabulation data analyses shows that there exist socio-technical aspects in all phases of the ontology engineering. The results showed the interdependency between the technical aspects (tasks and technology) and the social aspects (people and structure) in all phases of the ontology

IOP Conf. Series: Materials Science and Engineering **325** (2018) 012023 doi:10.1088/1757-899X/325/1/012023

engineering process, in accordance with the socio-technical model proposed by [7]. These findings are used as evidence to confirm Leavitt's theory in the context of socio-technical system. These findings also validate the initial model proposed in ontology engineering methodology.

Finally, we took a completely inductive approach, to determine other concepts in the context of life-cycle model which were synthesized from the dataset to confirm the life-cycle conceptual model of the initial socio-technical model. In the phase frequency distribution, we confirm the emergence of phase such as "planning," "analysis," "design," "implementation," and "evaluation" which confirms the proposed life-cycle model. The life-cycle model proposed is slightly different to the Waterfall model by [10] and [12].

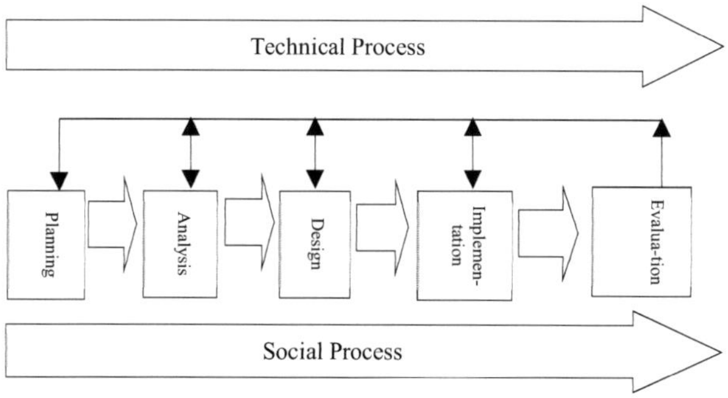

Figure 2. A Life-cycle Model in a Socio-technical Ontology Engineering

Validating the Model

Using a Cohen Kappa we initiated the calculation and found that the normal proportion of the total rating is 95% so that in this case there is a prevalence in the distribution according to the rating category. Table 2. below show an excerp from the calculation.

Table 2. Kappa coefficient calculation uses two rater data samples

		Rater 1		Row Margin	
		normal	abnormal		
Rater 2	normal	199	24	223	rm1
	abnormal	10	1	11	rm2
Column Margin		209	25	234	
		cm1	cm2		n

Therefore, there should some correction in the calculation of kappa to this prevalence. According [3] for the case of two rater the correction in Cohen Kappa, namely PABAK = 2Pa-1, wherein: raw percent of agreement (Pa)=200/232=86.2%, expected agreement (Pe)=87.1, prevalence index=(199-1)/234=0.85, PABAK=0.72. From the results of the calculation we obtained PABAK 0.72 which means that the resulting synthesis is reliable and the data is substantially valid. In the cases of observations made by more than two people, Kappa formula to calculate Pa can be used according to Fleiss. While a correction to the value of Kappa is done by using a calculation in [11], Pe is calculated using the correction 1/k, where k is the number of category. The results using multi-rater (three) are as follows: the validation is performed against every step and sequentially carried out starting from the translation in the first order, the translation into second order, then translation into category and the directly affected category, and lastly the synthesis into phase. Validation data in first-order translation, Pa=0.876, Pe=0.889, Kappa= -0121, adjusted Kappa against prevalence=0.752. Within second order translation we found, Pa=0.962, Pe=0.963, Kappa =-0.037, adjusted Kappa against prevalence = 0924. Validation of data in the synthesis into the first category: Pa=0.876, Pe=0.889, Kappa=-0.12, adjusted Kappa=0.752. Validation of data in the synthesis into second category: Pa=0.880, Pe = 0.886, Kappa=-0.058, adjusted Kappa=0.760. Validation of data in the synthesis into phase: Pa=0963, Pe=0.963, Kappa=-0.019, adjusted Kappa against prevalence=0.926.

5. Conclusion and Future Researches

This study highlights the complex interdependent relationship between technology factors and social factors and shows the need to address the socio-technical aspect in the ontology engineering methodology in order to have a balanced treatment between the technical process and the social process along the whole ontology engineering phases. The results of the analysis and synthesis are initial model which will be validated in a case study of an ontology development. The results of the case study will provide another validation of the model. Findings from our qualitative research synthesis will be refined through a survey on participants's perception on the operationalization of the methodology through case studies to empirically test the methodological model of the socio-technical ontology engineering. Overall the model validation is carried out by triangulation. The triangulation across various methods of data collection is very useful because it reinforce the validity of the model as well as providing multiple perspectives and to provide data and information that is complementary. In this paper the given description in the process is still in outline form (high-level). Improvements to this methodology will be done iteratively through case studies in ontology development which are undertaken to get a detailed picture of sub-processes and sub-activities.

Acknowledgement

The author would like to thank to Head of E-Government and E-Business laboratory in University of Indonesia and University of Indonesia as a funder of this research.

References

[1] Al-Debei M M, and Fitzgerald G 2009 OntoEng: a design method for ontology engineering in information systems *Proc. of the ACM OOPSLA'09 (ODiSE 2009)* pp 1-25.

[2] Britten N, Campbell R, Pope C, Donovan J, Morgan M, and Pill R 2002 Using meta ethnography to synthesise qualitative research: a worked example *J. of Health Services Research and Policy* **7(4)** 209-15.

[3] Byrt T, Bishop J, and Carlin J B 1993 Bias, prevalence and kappa *J. of Clinical Epidemiology* **46(5)** pp 423-9.

[4] Herrmann T, Loser K U, and Jahnke I 2007 Sociotechnical walkthrough: a means for knowledge integration *The Learning Organization* **14(5)** pp 450-64..

[5] Holsapple C W, and Joshi K D 2004 A formal knowledge management ontology: Conduct, activities, resources, and influences *J. of the Assoc. for Information Sci. and Technol.* **55(7)** pp 593-612.

[6] Kotis K, and Vouros G A 2006 Human-centered ontology engineering: The HCOME methodology *Knowledge and Information Syst.* **10(1)** pp 109-31.

[7] Leavitt H J 1965 Applying Organizational Change in Industry: Structural, Technological and Humanistic Approaches *J. G. March (Ed.) Handbook of Organizations* (Chicago, IL, USA: Rand McNally) pp 1144-70.

[8] Noblit, George W, and Hare R Dwight 1988 *Meta-Ethnography: Synthesizing Qualitative Studies* Vol. 11 (Sage).

[9] Pereira C, Sousa C, and Soares A L 2009 Short-term semantic consensus: towards agile ontology specification for collaborative networks *Working Conf. on Virtual Enterprises* (Berlin, Heidelberg: Springer) pp 301-10.

[10] Pressman R S 2010 *Software Engineering: A Practitioner's Approach* Ed. 7 (McGraw-Hill Higher Education).

[11] Randolph J J 2005 *Free-Marginal Multirater Kappa (multirater K [free]): An Alternative to*

Fleiss' Fixed-Marginal Multirater Kappa [Online Submission].
[12] Sommerville I 2011 *Software engineering* Ed. 9 (USA: Addison-Wesley).
[13] Suárez-Figueroa M C, Gómez-Pérez A, and Fernández-López M 2012 The NeOn methodology for ontology engineering *Ontology Eng. in a Networked World* (Berlin,Heidelberg: Springer) pp 9-34.
[14] Tempich C, Pinto S, Staab S, and Sure Y 2004 A case study in supporting DIstributed, Loosely-controlled and evolvInG Engineering of oNTologies (DILIGENT) *Proc. of the 4th Int. Conf. on Knowledge Management (I-KNOW'04)* pp 225-32.
[15] De Nicola A, Missikoff M, and Navigli R 2005 A proposal for a unified process for ontology building: UPON *Database and Expert Systems Applications* (Berlin/Heidelberg: Springer) pp 655-64.
[16] Pinto H S, Staab S, and Tempich C 2004 DILIGENT: Towards a fine-grained methodology for DIstributed, Loosely-controlled and evolvInG Engineering of oNTologies *Proc. of the 16th European Conf. on Artificial Intelligence* (IOS Pres) pp 393-7.
[17] Fernández-López M, Gómez-Pérez A, and Juristo N 1997 *Methontology: from ontological art towards ontological engineering.*
[18] Kotis K, and Vouros G A 2006 Human-centered ontology engineering: The HCOME methodology *Knowledge and Information Syst.* **10(1)** pp 109-31.
[19] Sure Y, Staab S, and Studer R 2002 Methodology for development and employment of ontology based knowledge management applications *ACM Sigmod Record* **31(4)** pp 18-23.

IOP Conf. Series: Materials Science and Engineering **325** (2018) 012024 doi:10.1088/1757-899X/325/1/012024

Refining the Relationships among Historical Figures by Implementing Inference Rules in SWRL

Nurul Fajrin Ariyani, Madis Saralita, Sarwosri, and Riyanarto Sarno

Informatics Department, Institut Teknologi Sepuluh Nopember, Surabaya, Indonesia

E-mail: nurulfajrin@if.its.ac.id, madis.saralita@gmail.com

Abstract. The biography of historical figures is often fascinating to be known. Everything about their character, work, invention, and personal life sometimes are presented in their biography. The social and family relationships among historical figures also put into concern, especially for political figures, heroes, kings or persons who have ever been ruled a monarchy in their past. Some biographies can be found in Wikipedia as articles. Most of the social and family relationship contents of these figures are not completely depicted due to a various article's contributors and sources. Fortunately, the missing relatives of a person might reside in the other figures' biography in different pages. Each `Wikipedia` article has metadata which represents its essential information of content. By processing the metadata obtained from `DBpedia` and composing the inferencing rules (in the form of ontology) to identify the relationships content, it can generate several new inferred facts that complement the existing relationships. This work proposes a methodology for finding missing relationships among historical figures using inference rules in an ontology. As a result, our method can present new facts about the relationships that absent in the existing Wikipedia articles.

1. Introduction

A historical figure is a person whose name is remembered for his or her action, influence, or involvement in particular historical events in the past. The biography of historical figures have various characters of information depends on their types. Some kings have had a few noble titles due to their rule as a king, queen, or emperor of a monarchy. However, some of other historical figures have labeled as heroes because of their heroic acts in the past events which affected the history nowadays. A historical figure is an evidence of a notable event that happened in particular time. Each figure has his or her own story and personal life which had involved the other figures based on his or her social and family relationships. The relationships among figures could depict the kind of historical event that had involved them in the past.

By knowing the family relationships among kings, we will able to build a complete genealogy of them which leads to the understanding of throne inheritance and the origin of the kingdoms. While by knowing the social relationships, we will know the alliance of a historical figure, what political parties he or she was involved and whether some figures were lived and cooperate in the same period of time. Both subjects are parts of history learning materials in schools. Thus, adding relationships among historical figures can give contributions to enrich knowledge learning materials for students.

The biography of historical figures often presents their social and family relationships with the other figures. A free online dictionary such as Wikipedia has preserved a large number of historical figure's biographies in the form of article pages. The relationships information that depicted in Wikipedia biographies may appear incomplete due to the different article's contributors or sources. Moreover, the name of historical figure tends to have multiple aliases which can burden the process of manually finding the missing relationships among them. Each Wikipedia article has metadata in the form of RDF format that stored in `DBpedia`. Fortunately, these missing relationships can be inferred from the other related pages by processing articles' metadata based on predefined inference rules that created in SWRL.

However, the inference rules reasoning new facts about relationships use specific terms that defined in an ontology. Thus, to be able to implement the rules, the original facts (existing Wikipedia metadata) need to be modeled using an ontology. An ontology defines `classes`, `properties`, and `instances` of a particular domain. Actor is one of the domains in cultural heritage ontology. The scope of actor involves a *person*, *group*, and *organization*. Historical figures including kings and heroes are categorized in *person* agent.

An ontology was built for specific purposes, and only valid within its domain. To create an ontology, we shall modify or integrate several existing ontologies rather than built it from scratch [1]. Several studies about building metadata of cultural heritage, including person were conducted by adopting CIDOC-CRM ontology which focuses on the relation between event, temporal, and spatial [2,3]. However, to find the relationships among historical figures, we need to combine several ontologies that related to biography and friendship relations. Our proposed ontology was developed in *Ontology Web Language (OWL)* using Protégé as the ontology code editor tool.

2. Related Works

In the previous works, a *Biography Light Ontology* (BLO) was built by composing several well-known ontologies such as *ABC Ontology*, *Linking Open Description of Events Ontology* (LODE), *Friend of a Friend* (FOAF), *BIO Vocabulary*, *Relationship Vocabulary*, CIDOC-CRM and *Bibliographic Ontology* (Bibo). This ontology's structure is more dynamic to represent biographical events rather than other models [4]. However, it lacks some specific terms regarding to describe a group, family, and organization representation. Its interoperability between the existing specifications and relevant standards for biography encoding still need improvement.

The other ontology, `LODE`, provides comprehensive evaluations of event ontology and Interlingua model to enhance the interoperability among the existing standards. LODE is an *event-centric* ontology which relates some *instances* based on their involvement in the same single event [5]. But in fact, the socio family relationships among historical figures that represented in DBpedia biography were not complemented with information about the figure's participation in an event, even though some specific events may explicitly appear in the metadata.

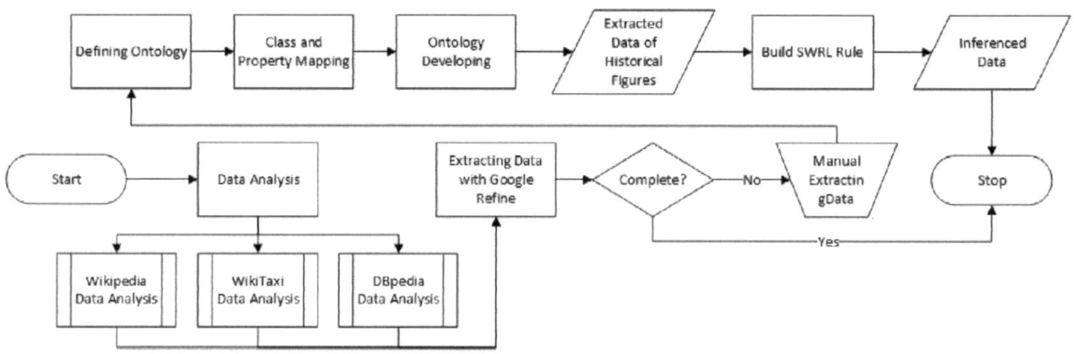

Figure 1. The Process Diagram

`FOAF` is a project intended for connecting between a person and his or her information via the web. FOAF is more suitable to depict a network between people in the modern life which has been influenced by social media technology [6]. However, FOAF is still used practically as a fundamental ontology and often combined with other well-known ontologies such as Bibo and DBpedia [7].

One of the other commonly used biography ontology named `BIO`. BIO document consists of a vocabulary to model biography information about a person and his or her background [8]. BIO provides a `class` for *event* and `properties` to relate a *person* to *event* and *event* to its *place* and *time* [9]. BIO defines several relationships such as *Father, Mother, Child, Birth, Death,* and so forth. But it is not sufficient to describe other family relationships outside the primary family member. To complement the relationship properties in the family domain, we used `FamilyTree` that was derived from University of Manchester's portal[1]. FamilyTree is one of the complex and comprehensive ontologies in the domain of family. The main purpose of this ontology is to generate new facts by minimizing relationships (property usage) and optimizing inference (rules). However, to enhance the inference capability, we added several SWRL rules to support the axioms that had been decided in the Family Tree. SWRL rules have flexibility in representing rule facts in common condition which including text and math operations also can be implemented in many cases other than genealogy [10].

Several works related to modeling the genealogy or family relationships using SWR rules to enriching the knowledge have done for many purposes, mostly for medical issues. Paper [11] employed GenOnto to structure the relatives of a patient that tend to suffer from Li-Fraumeni Syndrome. Its GenOnto covers until the third degree of relatives, i.e., cousins. An application for helping elderly to remember their family, called CAPTAIN MEMO, also implemented an ontology and inference rules to represent family relationships [12]. The genealogy domain is suitable for defining the real world semantics as depicted in [13]. In this paper, the authors implemented several axioms in their reference domain to determine the memberships of its classes. Based on this works, we can conclude that inference rules contribute in enriching the family relationships by producing inferred facts from existing axioms.

The ontology proposed in this work merged FamilyTree, FOAF, and BIO by performing the classes and properties combination. The merged ontology was developed by adding several new classes and properties to facilitate terms that could not be accommodated by the three original ontologies.

3. Methodology

The merger and development of the ontology were aimed to produce an ontology that can define the relationship type among person. Figure 1 shows the process diagram implemented in this work. Each process is described in the next following paragraph.

3.1. Data Analysis

Data analysis proceeded by taking several metadata of historical figures from DBpedia. To build a complete ontology that covers all things of the domain needs high cost due to the massiveness of data, time processing, and the inevitably changing facts over times [14]. Thus, some historical figures (`person`) that observed as data are people who considered of having a lot of relationships with others. Each `person` has biography related to `place` (e.g., *BirthPlace* and *DeathPlace*), `time` (e.g., *BirthDate* and *DeathDate*), and `event` (e.g., *Perang_Aceh*, *Proklamasi*, etc).

At first, we tried to create a metadata manually from Wikipedia pages. The label and term produced in this process did not have any standardization which would be difficult to merge in the later phase. Next, we utilized WikiTaxi to catch the XML format of Wikipedia article. We found that this XML also could not be processed efficiently since its taggings were not structured in order. Due to this reasons, we decided to get the metadata from DBpedia comfortably.

[1] http://www.co-ode.org/roberts/family-tree.owl

DBpedia provides the structure and content extraction (metadata) of Wikipedia articles. In fact, the metadata is composed of triples which describe any essential context on a web page as written below:

(*Subject, Predicate, Object*)

`Subject` represents any kinds of resources that exist in a world wide web and can be identified by its URI (Universal Resource Identifier). A `Predicate` links between subject and `object` in a particular context, but only limited to terms that listed in the vocabulary (ontology). It may have roles as an *object property* or a *data property*. A predicate that links between two URIs called object property. Otherwise, a predicate that links a URI with a literal value named data property. For example, there is a fact mentioned that Cut Nyak Dhien is the spouse of Teuku Umar. In DBpedia, this fact will be transformed into a triple as below.

(*CutNyakDhien, dbpedia–owl:spouse, TeukuUmar*)

Table 1. The List of Imported Classes and Properties

BIO	
Classes	Agent, Event, Group Event, Marriage, Individual Event, Birth, Burial, Death, Graduation, Person, Wedding, Event Generic
Object Properties	differentFrom, Agent, Partner, Principal
FamilyTree	
Classes	Man, Woman, Sex, Male, Female
Object Properties	hasChild, hasSon, hasDaughter, hasParent, hasFather, hasMother, hasHusband, hasWife, isChildOf, isSonOf, isDaughterOf, isParentOf, isFatherOf, isMotherOf, inLawOf, isSpouseOf, isHusbandOf, isWifeOf
Data Properties	hasName, hasFirst GivenName, hasFamilyName, knownAs, alsoKnownAs, formerlyKnownAs, hasEventYear, hasBirthYear, hasDeathYear

3.2. Ontology Developing
In the ontology design, we merged three different kind ontologies which are FamilyTree, FOAF, and BIO. The merged ontology, which named `iHero`, later has been adjusted to preserve some information of the biography's metadata. The terms which came from two different ontologies but have a similar domain could be integrated as a contextual information. However, if there is still any information that could not be addressed by the integrated ontology, we may add new terms to refine it [15].

3.3. Importing Classes and Properties
The classes and properties of iHero were carefully chosen from three original ontologies by considering the context of each term that usually described in the annotation. To ensure that the related terms have the same meaning, we also utilized `SUMO`, a credible upper ontology that widely used in the ontology merging [16]. The list of classes and properties with its originality is shown in Table 1.

3.4. Equivalent and Invers
The equivalent classes are two or more classes which have exactly same meaning or context. This equivalent attribute was derived from the original ontology. In iHero ontology, *Mariage* class is equivalent to *WeddingEvent Generic* class. Meanwhile, for its properties, *hasChild* property is equivalent to *isParentOf* property, and *hasParent* is equivalent to *isChildOf*. For example, if there is a fact about Cut Gambang `hasParent` Teuku Umar, it also means that Cut Gambang `isChildOf` Teuku Umar.

Table 2. The SWRL Rules in the Ontology

No	Property	Rule
R1	*isSpouseOf*	$Marriage(m) \wedge Female(sf) \wedge Male(sm) \wedge Partner(m,sf) \wedge Partner(s,sm) \rightarrow isSpouseOf(sm,sf)$
R2	*hasHusband*	$Mariage(m) \wedge Female(sf) \wedge Male(sm) \wedge Partner(m,sf) \wedge Partner(m,sm) \rightarrow isSpouseOf(sm,sf)$
R3	*hasWife*	$Marriage(m) \wedge Female(sf) \wedge Male(sm) \wedge Partner(m,sf) \wedge Partner(m,sm) \rightarrow hasWife(sm,sf)$
R4	*hasFather*	$Male(m) \wedge Person(p) \wedge hasParent(p,m) \rightarrow hasFather(p,m)$
R5	*hasMother*	$Female(f) \wedge Person(p) \wedge hasParent(p,f) \rightarrow hasMother(p,f)$
R6	*hasChild*	$Female(f) \wedge Male(m) \wedge Person(p) \wedge hasParent(p,f) \wedge hasParent(p,m) \rightarrow hasChild(f,p) \wedge hasChild(m,p)$
R7	*hasSon*	$Male(m) \wedge Person(p) \wedge haschild(p,m) \rightarrow hasSon(p,m)$
R8	*hasDaughter*	$Female(f) \wedge Person(p) \wedge hasChild(p,f) \rightarrow hasDaughter(p,f)$
R9	*hasSibling*	$Female(f) \wedge Male(m) \wedge Person(p) \wedge Person(q) \wedge hasFather(p,m) \wedge hasFather(q,m) \wedge hasMother(p,f) \wedge hasMother(q,f) \wedge DifferentFrom(p,q) \rightarrow hasSibling(p,q)$
R10	*hasBrother*	$Male(m) \wedge Person(p) \wedge hasSibling(p,m) \rightarrow hasBrother(p,m)$
R11	*hasSister*	$Female(f) \wedge Person(p) \wedge hasSibling(p,f) \rightarrow hasSister(p,f)$
R12	*hasStepParent*	$Person(c) \wedge Person(p1) \wedge Person(p2) \wedge Person(p3) \wedge hasHusband(p1,p2) \wedge hasHusband(p1,p3) \wedge hasParent(c,p1) \wedge hasParent(c,p2) \wedge DifferentFrom(p2,p3) \rightarrow hasStepParent(c,p3)$
R13	*hasStepFather*	$Male(m) \wedge Person(p) \wedge hasStepParent(p,m) \rightarrow hasStepFather(p,m)$
R14	*hasStepMother*	$Female(f) \wedge Person(p) \wedge hasStepParent(p,f) \rightarrow hasStepMother(p,f)$
R15	*hasStepSon*	$Male(m) \wedge Person(p) \wedge hasStepChild(p,m) \rightarrow hasStepSon(p,m)$
R16	*hasStepDaughter*	$Female(f) \wedge Person(p) \wedge hasStepChild(p,f) \rightarrow hasStepDaughter(p,f)$
R17	*hasStepSibling*	$Person(c1) \wedge Person(c2) \wedge Person(p) \wedge hasChild(p,c2) \wedge hasStepParent(c1,p) \rightarrow hasStepSibling(c1,c2)$
R18	*hasStepBrother*	$Male(m) \wedge Person(p) \wedge hasStepSibling(p,m) \rightarrow hasStepBrother(p,m)$
R19	*hasStepSister*	$Female(f) \wedge Person(p) \wedge hasStepSibling(p,f) \rightarrow hasStepSister(p,f)$
R20	*isAllyOf*	$Event(w) \wedge Group(g) \wedge Person(p1) \wedge Person(p2) \wedge hasAllegiance(p1,g) \wedge hasAllegiance(p2,g) \wedge hasCommander(w,p1) \wedge hasCommander(w,p2) \wedge DifferentFrom(p1,p2) \rightarrow isAllyOf(p1,p2)$
R21	*isEnemyOf*	$Event(w) \wedge Group(g1) \wedge Group(g2) \wedge Person(p1) \wedge Person(p2) \wedge hasAllegiance(p1,g1) \wedge hasAllegiance(p2,g2) \wedge hasCommander(w,p1) \wedge hasCommander(w,p2) \wedge DifferentFrom(p1,p2) \rightarrow isEnemyOf(p1,p2)$

Same as equivalent attribute, an inverse one also can be attached to classes or properties. However, it has an opposite meaning with the equivalent attribute. For example, *hasChild* property is an inverse of *hasParent* property. Thus, if there is fact about Cut Nyak Dhien `hasChild` Cut Gambang, it also means that Cut Gambang `hasParent` Cut Nyak Dhien.

3.5. Build SWRL Rules

The purposes of designed rules are to find relationships between one person to others. Rules were built in SWRL (Semantic Web Rule Language) by composing a pair of *antecedent* and *consequent* [17]. If the value of antecedent is true, then its consequence's value will also be true. The rules were expected to be able to create new facts related to a figure that is still absent in DBpedia by doing inferencing process.

We created 21 rules to find social family relations, as mentioned in Table 2. Please note that there is no automatic way to generate the inference rules; hence they need to be built by hand by code the family genealogy rules in SWRL. Each of the family relationship rules was derived from several family history term in [18].

For example, the rule *R*21 will be executed by utilizing class *Event, Group,* and *Person*. A single *Event*(*w*) which is probably an event of war, two groups *Group*(*g*1) and *Group*(*g*2) which each represents a nation, and two people *Person*(*p*1) and *Person*(*p*2) are involved in this rule as mentioned in Table 2. $Person(p1)$ has devoted to nation $Group(g1)$, and otherwise $Person(p2)$ has devoted to nation $Group(g2)$. Both people have participated as the commander in the same war $Event(w)$. However, $Person(p1)$ and $Person(p2)$ are different individuals. Also, they both have devoted to different $Group$. If all of this conditions are met in the existing facts, then it will produce a new information that $Person(p1)$ is the enemy of $Person(p2)$.

4. Evaluation

We used 20 meta data of historical figure's biographies as our experimental data regarding to their completeness of properties in DBpedia. These 20 figures are: Soekarno, Fatmawati, Mohammad Hatta, Hasyim Asyari, Mohammad Natsir, Cut Nyak Dhien, Teuku Umar, Cut Nyak Meutia, Teuku Nyak Arif, Pierre Tandean, Abdul Haris Nasution, Ahmad Yani, Ahmad Dahlan, Nyai Ahmad Dahlan, Tuanku Imam Bonjol, Wahid Hasyim, Mas Mansoer, Diponegoro, Sudirman, dan Slamet Rijadi. We have evaluated each of the 21 rules built in SWRL to ensure the correctness of its inferred facts derived from existing data. We applied Pellet Reasoner as the inferencing or reasoning engine and utilized Protégé to for editing and querying individuals. The complete results of evaluation are shown in Table 5.

The second column in Table 5 depicts one sample of the inferred facts resulted from reasoning, while in the third column is the correctness status of inferred facts that thoroughly compared with actual data from other sources. Several figures in second column are followed by their Wikipedia's article ID for clearness purposes. However, there are also some other figures that are not followed by IDs. It means that these figures were inferred from the reasoning process (not directly derived from a Wikipedia article).

For example, to evaluate the rule of `isEnemyOf` we compared between existing facts and inferred facts of Sudirman as shown in Table 3 and Figure 2. One of the inferred facts mentioned that Sudirman *isEnemyOf* William Franken as result of the rule in $R21$. William Franken *hasAllegiance* with Netherlands as this fact is explicitly written in Wiliam Franken's biography. Both Sudirman and William Franken were commander of Indonesian National Revolution event, though, they were not joined in the same allegiance. Due to this matters, the reasoner concluded that Sudirman *isEnemyOf* William Franken.

For example of family relationships rule, we put Cut Nyak Dhien biography as a sample. Table 4 shows several existing facts of this figure (please note that we only cropped properties related to family relations for clearness purposes). In Figure 3, it appeared that Cut Nyak Dhien *hasStepChild* Teuku Raja Sulaiman. This inferred fact resulted from *hasStepParent* rule in $R12$ and the inverse properties between *hasStepParent* and *hasStepChild.* Teuku Raja Sulaiman is the child of Teuku Umar and Cut Meuligou, Teuku Umar's wife from the previous marriage. Since Cut Meuligou and Cut Nyak Dhien are different individuals, the reasoner executed the rule and concluded that Teuku Raja Sulaiman *hasStepParent* Cut Nyak Dhien. Since *hasStepChild* is inverse of *hasStepParent* property, so it also produced that fact.

Table 3. Existing Metadata of Sudirman

Individuals	Sudirman	
Dbpedia resources	http://id.dbpedia.org/resource/Sudirman	
Types	foaf:Person, Hero, Male	
Object Property Assertions	**Property Name**	**Value**
	hasAllegiance	G_Indonesia
	isCommanderOf	E_Indonesian_National_Revolution
	isCommanderOf	'Agresi Militer Belanda I'
	isCommanderOf	E_Battle_of_Ambarawa
	isInfluencedOf	E_Tan_Malaka
	hasSubject	S_Indonesian_generals

Table 4. Existing Metadata of Cut Nyak Dhien

Individuals	Cut Nyak Dhien	
Dbpedia resources	http://id.dbpedia.org/resource/Cut_Nyak_Dhien	
Types	foaf:Person, Hero, Female	
Object Property Assertions	**Property Name**	**Value**
	hasSpouse	P_Teuku_Umar
	hasChild	P_Cut_Gambang
	isCommanderOf	E_Aceh_War
	knownFor	G_National_Heroine

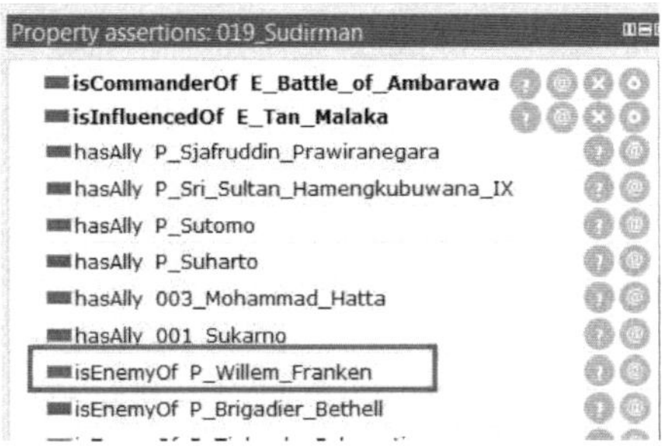

Figure 2. Inferred Facts of Sudirman Gathered from Protégé

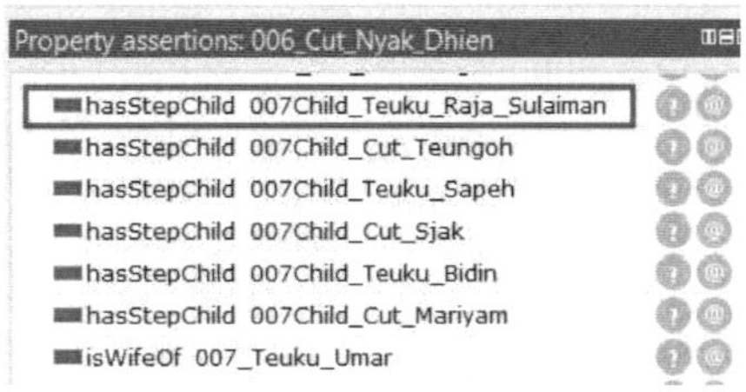

Figure 3. Inferred Facts of Cut Nyak Dhien Gathered from Protégé

Table 5. The Evaluation of 21 Rules

No.	Inferred Fact	Correct
R1	Cut_Nyak_Jauhari i isSpouseOf – Teuku_Nyak_Arif_6667132	v
R2	Cut_Nyak_Dhien_6786184 – hasHusband – Teuku_Umar_6688121	v
R3	Soekarno_6816787 – hasWife – Oetari_1836277	v
R4	Ade_Irma_Suryani_6238101 – hasFather – Abdul_Haris_Nasution_6666891	v
R5	Cut_Gambang - hasMother – Cut_Nyak_Dhien_6786184	v
R6	Mohammad_Hatta_6810795 - hasChild – Halida_Hatta_6566647	v
R7	Fatmawati_6627334 - hasSon – Guruh_Soekarnoputra_6657412	v
R8	Rahmi_Rachim_3188629 - hasDaughter – Meutia_Hatta_6772623	v
R9	Guruh_Soekarnoputra_6657412 - hasSibling – Megawati_Soekarnoputri_6596840	v
R10	Sukmawati_Soekarnoputri_6516890 - hasBrother – Guruh_Soekarnoputra_6657412	v
R11	Teuku_Bidin - hasSister – Cut_Teungoh	v
R12	Cut_Mariyam - hasStepParent – Cut_Nyak_Dhien_6786184	v
R13	Cut_Gambang - hasStepFather – Ibrahim_Lamnga	v
R14	Teuku_Raja_Sulaiman - hasStepMother – Cut_Nyak_Dhien_6786184	v
R15	Cut_Nyak_Dhien_6786184 - hasStepSon – Teuku_Raja_Sulaiman	v
R16	Cut_Nyak_Dhien_6786184 - hasStepDaughter – Cut_Teungoh	v
R17	Cut_Gambang - hasStepSibling – Teuku_Raja_Sulaiman	v
R18	Cut_Gambang - hasStepBrother – Teuku_Raja_Sulaiman	v
R19	Cut_Gambang - hasStepSister – Cut_Teungoh	v
R20	Sudirman_1800798 - isAllyOf – Mohammad_Hatta_6810795	v
R21	Diponegoro_6728234 - isEnemyOf – Hendrik_Merkus_de_Kock_6668025	v

As we see from Table 5, we may conclude that all of our inference rules were successfully implemented and could produce new inferred facts based on the rules. However, the `isAllyOf` and `isEnemyOf` properties need to be revised further since a figure might change his/her allegiance in the past. We consider employing the timeline sequence to clear the validity period of each inferred facts. Furthermore, the reasoning process seems to be not efficiently implemented since we need to maximize the heap size memory of Protégé to complete the whole process. By utilizing 2 GB of memory, the reasoning process was completed in less than a minute. However, the time needed to export all of the inferred axioms (so it will be able to be queried) was more than 20 minutes.

5. Conclusion

We have developed an ontology for finding relationships among historical figures by merging FOAF, BIO, and FamilyTree scheme and building SWRL rules to reason the missing relations. Rules' functionalities could produce several new inferred facts based on the existing facts in the 20 data figures. Limited memory used during processing affects the complexity of reasoning process, though its need a lot more time to export the inferred axioms into physical files so they could be queried using Protégé's SPARQL endpoint.

Regardless of the success of the proposed ontology, there is a lack of methods in the reasoning process. The reasoning processing time will grow exponentially to the complexity of defined rules in the ontology. Our future works aim at this challenge of efficiently reasoning metadata, so it will be appropriate to implement in the real web based application for searching relationships of historical figures.

Acknowledgments
This research funding was granted by Lembaga Penelitian dan Pengabdian Masyarakat, Institut Teknologi Sepuluh Nopember Surabaya as Penelitian Pemula Dana Lokal ITS 2017.

References

[1] Gruber, T.R., 1995. Toward principles for the design of ontologies used for knowledge sharing?. *International journal of human-computer studies*, **43(5-6)**, pp.907-928.
[2] Ariyani, N.F., Hanafi, M.B., Ginardi, H. and Saralita, M., 2016, October. Linking spatial information of cultural heritage metadata with geo linked open data. In *Information Technology Systems and Innovation (ICITSI), 2016 International Conference on* (pp. 1-5). IEEE.
[3] Ariyani, N.F. and Yuhana, U.L., 2015, November. Generating cultural heritage metadata as linked open data. In *Information Technology Systems and Innovation (ICITSI), 2015 International Conference on* (pp. 1-6). IEEE.
[4] Ramos, M.R., Biography Light Ontology: An Open Vocabulary For Encoding Biographic Texts.
[5] Shaw, R., Troncy, R. and Hardman, L., 2009. LODE: Linking Open Descriptions of Events. *ASWC*, **9**, pp.153167.
[6] Al-Mukhtar, M.M.A. and Al-Assafy, A.T.A., 2014. The implementation of foaf ontology for an academic social network. *International Journal of Science, Engineering and Computer Technology*, **4(1)**, p.10.
[7] Lehmann, J., Isele, R., Jakob, M., Jentzsch, A., Kontokostas, D., Mendes, P.N., Hellmann, S., Morsey, M., Van Kleef, P., Auer, S. and Bizer, C., 2015. DBpediaa large-scale, multilingual knowledge base extracted from Wikipedia. *Semantic Web*, **6(2)**, pp.167-195.
[8] Davis and Galbraith, D., BIO: A vocabulary for biographical information. [Online]. Available: http://vocab.org/bio/. [Accessed 06 January 2016].
[9] Larson, R., Bringing lives to light: Biography in context (2010).
[10] Hartanto, H.A., Sarno, R. and Ariyani, N.F., 2016, August. Linked warning criterion on ontology-based key performance indicators. *In Technology of Information and Communication (ISemantic), International Seminar on Application for* (pp. 211-216). IEEE.
[11] Budaruiche, R.M., Wassermann, R., Patrao, D.F. and Achatz, M.I., 2015, November. Li-Fraumeni Ontology: A Case Study of an Ontology for Knowledge Discovery in a Cancer Domain. In *Intelligent Systems (BRACIS), 2015 Brazilian Conference on* (pp. 157-162). IEEE.
[12] Herradi, N., Hamdi, F., Mtais, E., Ghorbel, F. and Soukane, A., 2015, October. PersonLink: an ontology representing family relationships for the CAPTAIN MEMO memory prosthesis. In *International Conference on Conceptual Modeling* (pp. 3-13). Springer, Cham.
[13] De Carvalho, V.A., Almeida, J.P.A. and Guizzardi, G., 2014, June. Using reference domain ontologies to define the real-world semantics of domain-specific languages. In *International Conference on Advanced Information Systems Engineering* (pp. 488-502). Springer, Cham.
[14] Wu, D. and Hkansson, A., 2014. A method of identifying ontology domain. *Procedia Computer Science*, **35**, pp.504-513.
[15] Simperl, E., Sarasua, C., Ungrangsi, R. and Brger, T., 2011. Ontology metadata for ontology

IOP Conf. Series: Materials Science and Engineering **325** (2018) 012024 doi:10.1088/1757-899X/325/1/012024

reuse. *International Journal of Metadata, Semantics and Ontologies*, **6(2)**, pp.126-145.
[16] Niles, I. and Pease, A., 2001, October. Towards a standard upper ontology. /it In Proceedings of the international conference on Formal Ontology in Information Systems-Volume 2001 (pp. 2-9). ACM.
[17] Oconnor, M., Knublauch, H., Tu, S. and Musen, M., 2005. Writing rules for the semantic web using SWRL and Jess. *Protégé With Rules WS, Madrid.*
[18] Watson, J., 1996. *Ordering the family: Genealogy as autobiographical pedigree.* na.

Time Series Neural Network Model for Part-of-Speech Tagging Indonesian Language

Theo Tanadi

School of Electrical Engineering and Informatics, Institut Teknologi Bandung

18214041@std.stei.itb.ac.id

Abstract. Part-of-speech tagging (POS tagging) is an important part in natural language processing. Many methods have been used to do this task, including neural network. This paper models a neural network that attempts to do POS tagging. A time series neural network is modelled to solve the problems that a basic neural network faces when attempting to do POS tagging. In order to enable the neural network to have text data input, the text data will get clustered first using Brown Clustering, resulting a binary dictionary that the neural network can use. To further the accuracy of the neural network, other features such as the POS tag, suffix, and affix of previous words would also be fed to the neural network.

Keyword: *Neural network, Part-of-speech tagging, Brown clustering*

1. Introduction

Part-of-speech (POS) tagging is one of the most basic natural language processing tasks. The main objective of POS tagging is to give a part-of-speech tag to each token in a given text. There are many approaches to solve this problem, such as rule-based tagger, Hidden Markov Model [1], and artificial neural network. The usage of artificial neural network in POS tagging task has been proven to perform better than other approaches, with many variations.

Feedforward neural network that has been modified has been used for POS tagging task since 1990s [2]. Other more sophisticated neural network architecture started to emerge along with many other advancements in natural language processing field. The direction of POS tagging task approach started to lean towards deep learning and recurrent neural network [3].

In this paper we will try to create a simple time series neural network model to solve POS tagging task. Time series neural network is a feedforward neural network which have previous output(s) as its input. Time series neural network is slightly different from recurrent neural network. Recurrent neural network has recurrent connection, making its history length unlimited, whereas time series neural network only can have its history length specified by the number of previous outputs that is going to be fed to the next input.

In the next section, we will discuss about the basic of artificial neural network and Brown clustering. After that, the model and training method of the neural network will be explained. Analysis and testing plan will follow. This paper will be finished with description of future work and conclusion.

2. Artificial Neural Network

Artificial neural network is a type of computing system that is inspired from brain neural network. Artificial neural network consists of multiple layer of neurons [4]. A basic artificial neural network has 2 layer, its input and its output. A more complex artificial neural network has another layer between its input layer and output layer. Each neuron that is connected to each other will have a weight that determines the strength of the connection. Each neuron also has an activation function that will transform the value of the input.

Artificial neural network works by feeding input to its neuron, multiply it with corresponding weight, transform it with its activation function, and transfers its output all the way through from its input layer to its output layer. With output as the output of the neuron, f as the activation function, input as the input of the neuron, and weight as the weight from the previous neuron, the output of a neuron can be represented by this formula:

$$output = f(input \cdot weight)$$

3. Brown Cluster

One of artificial neural network drawback is that it cannot process raw text. Some adjustment need to be made before feeding text into neural network. There are a few methods that can be used to enable neural network to accept text input, such as word embedding [5] and Brown clustering. The method that will be used in this paper is Brown clustering.

Brown clustering works by clustering words based on their occurrence and the occurrence of nearby words. The idea is that similar words will have similar distributions of their nearby words. Brown clustering inputs a large text corpus and outputs a binary dictionary of words that exist in the input corpus. Similar words will have similar most significant bit (MSB). To use the dictionary from Brown clustering, we take fixed length MSB from the dictionary to represent the words.

4. Model

In this paper, the neural network model that is designed to do POS tagging task will have 3 layer, which is one input layer, one hidden layer, and one output layer. Each neuron in the input layer will be connected to every neuron in the hidden layer and each neuron in the hidden layer will be connected to every neuron in the output layer. It is possible to use any other activation function, but in this paper each neuron will use sigmoid function as its activation function.

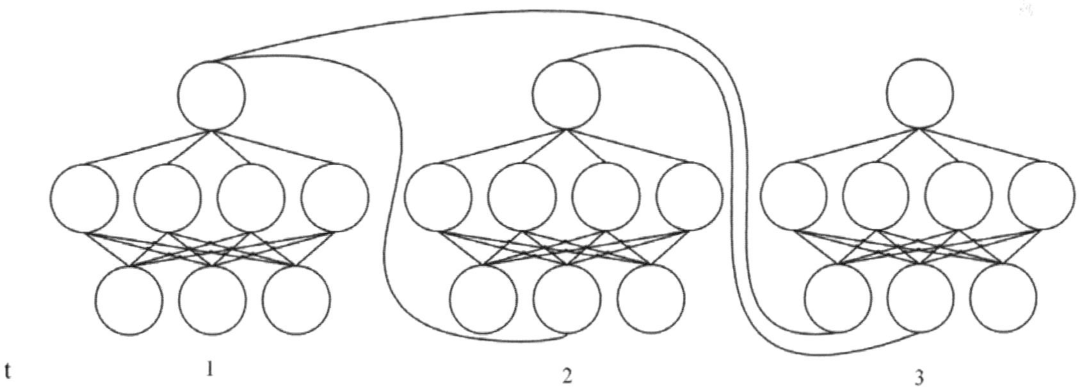

Figure 1. Proposed neural network model diagram (simplified). The bottom layer is the input layer, the middle layer is the hidden layer, and the top layer is the output layer. The numbers at the bottom represent the number of iteration the neural network have done.

There are several inputs that is going to be fed to the model. The first one is the word that is going to be tagged. The word will get transformed first according to the binary dictionary from the Brown

IOP Conf. Series: Materials Science and Engineering **325** (2018) 012025 doi:10.1088/1757-899X/325/1/012025

cluster. The size of the binary word hasn't been decided yet. The model also can get the POS tag from two previous words as its input. Another input that the model can have is the suffix and prefix from the word that is going to be tagged. The suffix and prefix of the word are received from CS stemmer [6]. The input layer is expected to have seventy to ninety nodes.

As for the hidden layer, the size of the node will be determined by trial and error [7]. But as a baseline, one hidden layer and the product of the number of nodes in input layer and output layer as the size of the hidden layer will be used.

For the output layer, the only output is the POS tag of the word. The POS output will be represented by N neuron where N is the number of POS tagset [8]. Currently, there is no standard Indonesian language corpus and tagset, but there are several previous works that attempt to create a standard Indonesian language tagset. This model will use proposed Indonesian tagset by Dinakaramani et al [9].

5. Model Training

The model use gradient descent method to find the optimal weight for each neuron. These are the steps for the step to train:

1. First the method will try to calculate the output using the initial weight. Error for the output layer is calculated by using this formula:

$$output\_error = desired\_output - calculated\_output$$

2. After the error is found, we calculate the delta for the output layer. With $f(x)$ as the activation function and x as the calculated output, the delta for the output layer is calculated by using this formula:

$$output\_delta = output\_error * \frac{d\,f(x)}{dx}$$

By using sigmoid function as our activation function, the delta formula can also be written as:

$$output\_delta = output\_error * (calculated\_output * (1 - calculated\_output))$$

3. We calculate the error for the hidden layer by back propagating the error from the output layer to the hidden layer. The error from the output layer is called confidence weighted error since it is calculated only from the desired output and the error from the hidden layer is called contributed weighted error since it is calculated from the error from the output layer. The error for the hidden layer is calculated by using this formula:

$$hidden\_error = output\_delta \cdot hidden\_to\_output\_weight$$

4. With $calculated\_hidden$ as the value of the hidden neuron when the output is calculated, the delta from the hidden layer with this formula:

$$hidden\_delta = hidden\_error * (calculated\_hidden * (1 - calculated\_hidden))$$

5. The new weight of the hidden layer to the output layer is calculated with this formula:

$$hidden\_to\_output\_weight = hidden\_to\_output\_weight + (calculated\_hidden \cdot output\_delta)$$

6. With $input$ as the input value that is fed to the input neuron, the new weight of the input layer to the hidden layer is calculated with this formula:

$$input\_to\_hidden\_weight = input\_to\_hidden\_weight + (input \cdot hidden\_delta)$$

6. Analysis and Testing Plan

There are some factors that can manipulate the accuracy of the neural network, such as the input features, the size and number of hidden layer, and the neural network optimization algorithm. These factors can also affect the accuracy of the neural network that we made.

IOP Conf. Series: Materials Science and Engineering **325** (2018) 012025 doi:10.1088/1757-899X/325/1/012025

Previous works that uses feedforward neural network usually use the POS tag of some of previous words and the POS tag of some next words. The usage of the prefix and suffix of the word hasn't been done before. Words with certain combinations of prefix and suffix tend to have certain POS tag. For example, *permainan* (game), *pelarian* (runaway), and *perlawanan* (resistance) have "*pe-*" prefix, "*-an*" suffix and NN (noun) as its POS tag. Another example is *bermain* (playing), *berlari* (running), and *berdiri* (standing) have "*be-*" prefix, no suffix, and VB (verb) as its POS tag. There are some exceptions for these pattern, but it still might lead to better accuracy for the neural network. Testing need to be done to ensure if it will really increase the accuracy of the neural network.

There are many discussions about the number of hidden layer and the size of the hidden layer that are needed in a neural network architecture to achieve optimal results. Many of them agree that more hidden layer will lead to better accuracy especially for difficult object, but the universal approximation theory claims that a feedforward neural network with one hidden layer with finite number of hidden unit and arbitrary activation function is capable to approximate a finite set [10,11]. Trial and error is needed to determine the best hidden layer number and hidden layer nodes number for a neural network to do certain task.

Neural network optimization algorithm is the algorithm that is used by neural network to find the most optimal weights of each connection in the neural network. The weights are considered optimal when its loss function is as minimal as possible. There are many optimization algorithms that can be used in neural network such as gradient descent, genetic algorithm [12], Adam [13], and Adagrad [14]. This model will use gradient descent optimization algorithm since it is the simplest. It does not rule out the possibility of using another optimization algorithm since these optimization algorithm is architecture independent.

The biggest obstacle to implement this model is the lack of training data. There are several researches in natural language processing that uses Indonesian language, which the training data can be used for this model. With these existing training data, we can create binary dictionary of words using Brown clustering and train the model itself. Unfortunately, we have not get our hand into any of these data.

There are several matrices that can be used to compare the model that we created with other model from previous works, such as accuracy and speed. This model is expected to have same or better accuracy than the HMM method as the HMM method is one of the most popular method to do POS tagging task.

7. Future Works

One of the most important thing that needs to be done is to get training data for this model. The training data can be obtained from other pervious researches or by creating one our self. The more training data that can be obtained, the better. Training data that consists at least ten thousand words and their corresponding POS tag is expected. Some preprocessing might be needed beforehand so the data can be fed to the neural network.

The prototype of the neural network has been made. This prototype can only process predefined binary data. This prototype achieved 100% accuracy for a really small training data and test data. Some adjustment is needed for this prototype so this prototype become able to do its intended purpose with sufficient accuracy. This paper is created in relevance to knowledge management research that has been conducted previously [15,16].

8. Conclusion

We have designed a neural network model to do part-of-speech tagging task. The model has not implemented yet because the lack of training data. The model will be implemented in the future after acquiring sufficient training data.

Acknowledgement
This study is financially supported by Institut Teknologi Bandung (ITB) under Program Penelitian, Pengabdian kepada Masyarakat, dan Inovasi (P3MI). We would like to thank Krisdanto Surendro (Information System Laboratory – SEEI ITB) for giving guidance throughout the writing of this paper.

References

[1] Wicaksono A F and Purwarianti A 2010 HMM Based Part-of-Speech Tagger for Bahasa Indonesia *4th International MALINDO (Malay and Indonesian Language) Workshop.*
[2] Schmid H 1994 Part-of-Speech Tagging with Neural Networks *COLING 94 Proc. 15th Int. Conf. Comput. Linguist.* **1** p 5.
[3] Plank B, Søgaard A and Goldberg Y 2016 Multilingual Part-of-Speech Tagging with Bidirectional Long Short-Term Memory Models and Auxiliary Loss pp 412–8.
[4] Wang S-C 2003 Artificial Neural Network *Interdiscip. Comput. Java Program* 81–100.
[5] Mikolov T, Chen K, Corrado G and Dean J 2000 Distributed Representations of Words and Phrases and their Compositionality *CrossRef List. Deleted DOIs* **1** pp 1–9.
[6] Asian J 2007 Effective Techniques for Indonesian Text Retrieval *Thesis.*
[7] Stathakis D 2009 How many hidden layers and nodes? *Int. J. Remote Sens.* **30** pp 2133–47.
[8] Aly M 2005 Survey on Multiclass Classification Methods Extensible algorithms *Neural Networks* pp 1–9.
[9] Dinakaramani A, Rashel F, Luthfi A and Manurung R 2014 Designing an Indonesian part of speech tagset and manually tagged Indonesian corpus *Proc. of the Int. Conf. on Asian Language Processing 2014 (IALP 2014)* pp 66–9.
[10] Csáji B 2001 *Approximation with artificial neural networks.*
[11] Hornik K 1991 Approximation capabilities of multilayer feedforward networks *Neural Networks* **4** pp 251–7.
[12] McCall J 2005 Genetic algorithms for modelling and optimisation *J. Comput. Appl. Math.* **184** pp 205–22.
[13] Kingma D P and Ba J L 2015 Adam: a Method for Stochastic Optimization *Int. Conf. Learn. Represent. 2015* pp 1–15.
[14] Duchi J, Hazan E and Singer Y 2011 Adaptive Subgradient Methods for Online Learning and Stochastic Optimization *J. Mach. Learn. Res.* **12** pp 2121–59.
[15] Aradea, Suwardi I S and Surendro K 2015 An overview of multi agent system approach in knowledge management model *2014 Int. Conf. Inf. Technol. Syst. Innov. (ICITSI 2014) - Proc.* pp 62–9.
[16] Atrinawati L H and Surendro K 2009 Assessment for knowledge management readiness *Proc. 2009 Int. Conf. Electr. Eng. Informatics, (ICEEI 2009)* **2** pp 399–404

IOP Conf. Series: Materials Science and Engineering **325** (2018) 012026 doi:10.1088/1757-899X/325/1/012026

Medical Named Entity Recognition for Indonesian Language Using Word Representations

Arief Rahman[1]

[1]School of Electrical Engineering and Informatics (SEEI), Institut Teknologi Bandung, Indonesia

Email: [1]23516008@std.stei.itb.ac.id

Abstract. Nowadays, Named Entity Recognition (NER) system is used in medical texts to obtain important medical information, like diseases, symptoms, and drugs. While most NER systems are applied to formal medical texts, informal ones like those from social media (also called semi-formal texts) are starting to get recognition as a gold mine for medical information. We propose a theoretical Named Entity Recognition (NER) model for semi-formal medical texts in our medical knowledge management system by comparing two kinds of word representations: cluster-based word representation and distributed representation.

1. Introduction

The study of using NLP in the biomedical domain has been in the spotlight for the last few years. Its popularity soars mainly because of the abundance of information and the computing power available for processing information. Named Entity Recognition (NER) is an important NLP task that allows us to get important medical information from medical texts easily, which can be useful to get new knowledge (e.g. in indigenous systems [1]).

NER is a task of finding named entities (NE), which can be words and phrases, which belong to the same semantic classes. In the biomedical domain, the widely accepted named entities are diseases, signs/symptoms, anatomical signs, and drugs. Several NER tools have been developed for getting medical named entities, such as MedLEE [2], MetaMap [3] and cTAKES [4]. These tools use a rule-based system, which relies on comprehensive medical knowledge and vocabularies.

Currently, most state-of-the-art medical NER systems use machine learning models. NER in particular uses supervised machine learning, in which labeled data are used to train models and then used to recognize new, unlabeled data. In machine learning, NER is approached as a sequence labeling problem that aims to find the best label sequence (e.g. IO, BIO, and BMEWO format label) in a given sentence (sequence of words). Three of the most-used machine learning algorithms in medical NER are Conditional Random Field (CRF) [5], Maximum Entropy (ME), and Support Vector Machines (SVM) [6]. One big downside, however, is the fact that the supervised machine learning algorithms only works well when the features are manually extracted through feature engineering.

Typical NER systems mostly used orthographic features (e.g., capitalization of letters, prefix, and suffix), syntactic features (e.g. POS tags), and n-gram features. However, researchers found that the performance of supervised machine learning model can be improved by using "unsupervised" features, which usually replaces the word form feature as an alternative word representation. These alternative word representations are learned by using unlabeled corpus. Some of the most famous unsupervised features are Brown cluster [7] and word embedding [8][9]. Brown cluster is one of the word representations based on clustering. Word embedding, on the other hand, is a type of distributed word representation.

Most medical NER systems focused on formal medical texts, which contain formal linguistic styles. This makes it easier to find named entities because it is guaranteed to follow certain grammar. However, medical texts nowadays can also come from social media. They are usually found in informal medical chats or groups. These texts most of the time have both formal and informal grammar, which can be a big challenge for most NER systems. cTAKES [4] tried to process medical free-texts, with a satisfying F1-score (81,24%). However, the NER model in this system used a terminology-agnostic dictionary look-up algorithm.

In this research, we propose a theoretical model to our medical NER system by using two kinds of unsupervised features, which are cluster-based word representation and distributed word representation. In addition to its easy-to-use property, both of these features are also reliable to use on texts with informal language styles. The usage of word representation is expected to help in making a NER model that's easy to train, yet good at classifying named entities. In the following sections, we summarize the basic concepts of the Brown cluster and word embedding. We also explain our proposed theoretical model and the proposed experiments to be done. This study is conducted as a part of the Knowledge Management System (KMS) study conducted in [1] and [10].

2. Word Representations

In this section, we discuss two types of word representations: cluster-based word representation and distributed word representation. We do not discuss the distributional word representations because of its lack of development in the recent years.

2.1. Cluster-based Word Representation

Cluster-based word representation is a word representation that is obtained by inducing a clustering over words. There are two types of cluster-based word representation: hard clustering (Brown cluster) and soft clustering (HMM word representation).

Brown et al. [7] proposed a greedy agglomerative hierarchical clustering procedure that groups words to maximize the mutual information of bigrams. In this procedure, clusters are initialized with a single word in each cluster. After that, the clusters are merged greedily according to the bi-gram mutual information criterion. The resulting clusters are merged until only one big cluster remains.

The result of Brown cluster procedure is a low-dimensional representation of the vocabulary in the form of a hierarchical, binary tree. The low-dimensional representation mitigates the feature sparseness problem. The procedure runs in time complexity of $O(VK^2)$, with V being the size of the vocabulary and K being the number of clusters. Figure 1 shows an example of the Brown clustering result.

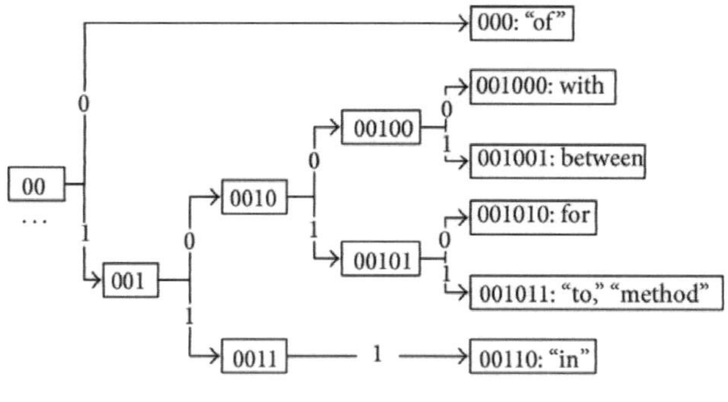

Figure 1. Proposed named-entity recognition system process [11].

Several studies have proved that this unsupervised clustering procedure has helped in increasing NER performances [12]. The hierarchical nature of the Brown clusters allows words to be represented at different levels of the hierarchy. The Brown cluster has one major downside: it does not consider wider contexts and word usage in general, using only bigram statistics.

2.2. Distributed Word Representation

Distributed word representation is dense, low-dimensional, and real-valued vector that represents its syntactic (and potentially semantic) properties. This representation is also called word embedding. Each of its dimension represents a latent feature of the word, which captures syntactic and semantic properties for the computer to understand. A distributed representation is compact and can represent an exponential number of clusters in a limited number of dimensions.

Word embeddings are usually made using neural language model, which uses neural networks as the underlying predictive model [13]. It is different from clustering-based word representations, in which they are usually induced under n-gram language models. Historically, training and testing of word embedding models have been slow, scaling to the size of the vocabulary [14]. Many studies to allow word embedding scaling to very large training corpora were done in recent years. The biggest breakthrough comes from Mikolov [8][9], which induces word representations using a neural network that has a log-linear complexity. Mikolov's design allows researchers to induce distributed word representation with much more efficiency without significant accuracy defect.

Mikolov proposed two architectures for efficient neural network language model (NNLM). The first one is called the Continuous Bag-of-Words model (CBOW) [8]. It uses a feedforward NNLM, with no non-linear hidden layer and shared projection layer for all words. The architecture uses both words from the history and the future (previous n tokens and next n tokens). The *bag-of-words* name comes from the fact that the position of the words in the window does not matter in the long run to determine the embedding.

The second architecture is called the skip gram model [8]. The skip gram architecture uses a different approach from CBOW, by maximizing the classification probability of a word based on another word in the same sentence. It does this by using each current word as an input to a log-linear classifier with continuous projection layer and predict words within a certain range before and after the current word. Increasing the context range improves both the quality of the resulting word vectors and the computational complexity. A weight decay penalty is given to distant words by doing less sampling in the training examples because they are usually less related to the current word.

Figure 2 shows both the CBOW and skip gram neural network architecture.

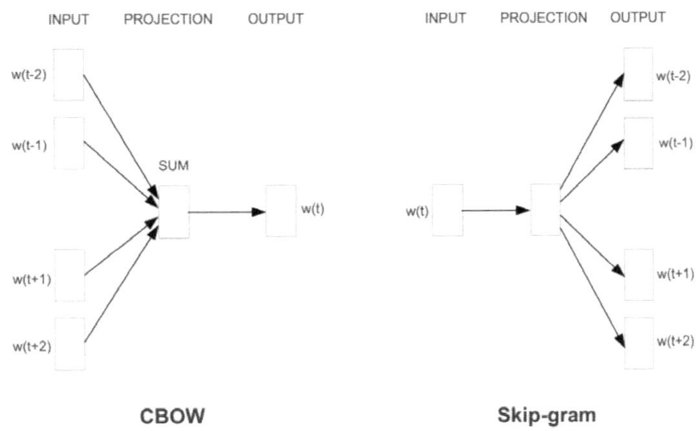

Figure 2. CBOW and skip gram architecture [8].

IOP Conf. Series: Materials Science and Engineering **325** (2018) 012026 doi:10.1088/1757-899X/325/1/012026

3. Proposed Named Entity Recognition Model

We propose the following NER model. First, we will crawl one or more medical-oriented social media groups to obtain medical semi-formal texts. After that, we use a language formalization tool to reduce the number of features and allow easier processing in the next steps. In the next step, we use will the final dataset for two purposes: train a word representation and create a NER dataset. For the first purpose, we use an unsupervised word representation training algorithm to train the word representation. For the second purpose, currently, we plan to annotate the dataset manually with named medical entities. Finally, we use both the NER dataset and the trained word representations to train a model using the artificial neural network algorithm. The whole process can be seen in Figure 3.

In the next subsections, we explain the important parts in our NER system, in which we will discuss the dataset, the word representations, and the model learning.

3.1. Data Sets

For this work, we will use the medical semi-formal texts fetched from several Facebook's Indonesian medical discussion groups. These groups use semi-formal language style medical texts, with texts ranging from medical information to medical questions. We will extract these texts using NodeXL. We plan to use around one year worth of medical semi-formal texts to use.

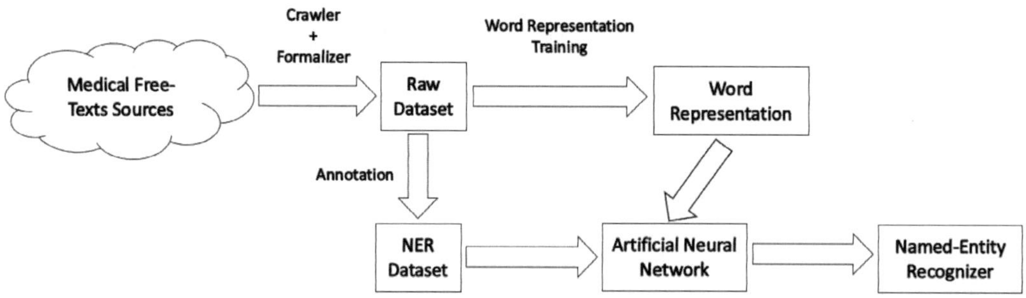

Figure 3. Proposed named-entity recognition model process.

To allow easier processing, we will formalize the texts obtained from the media social using our own rule-based Indonesian formalization tool. We will also delete all texts that don't contain any medical named entities. To detect these named entities, we will use a gazetteer that contains the most common medical terms in medical texts in general. We will manually tag the named entities on the whole texts after the whole processing steps are done.

3.2. Word Representation Features

In this work, we will use several word representations as features for our NER model. We will use the Brown Cluster [7] and Mikolov's word embedding [8]. To train these word representations, we use both the Indonesian Wikipedia dump[1] and a corpus containing formalized raw texts from several Facebook's Indonesian medical discussion groups. We plan to use around one year worth of medical semi-formal texts to use. We use these corpora to train both the formal and informal text styles.

For the word representation generator, we will use Percy Liang's Brown cluster generator[2] , which is implemented in Python and is easy to use and parameterize. We will also use Mikolov's word2vec[3] to generate the word embedding and use both the CBOW and skip-gram models.

[1] https://dumps.wikimedia.org/
[2] https://github.com/percyliang/brown-cluster
[3] https://code.google.com/archive/p/word2vec/

3.3. Neural Network Model and Word Representation Features

Our NER models will use the neural network, which has been used recently, with satisfying performance. Our network consists of an input layer, two hidden layers with ReLU activation function (with 100 nodes per layer), and a softmax output layer, which represents the probability of each named entity (including its BIO label) being the correct label. Figure 4 shows our proposed neural network architecture.

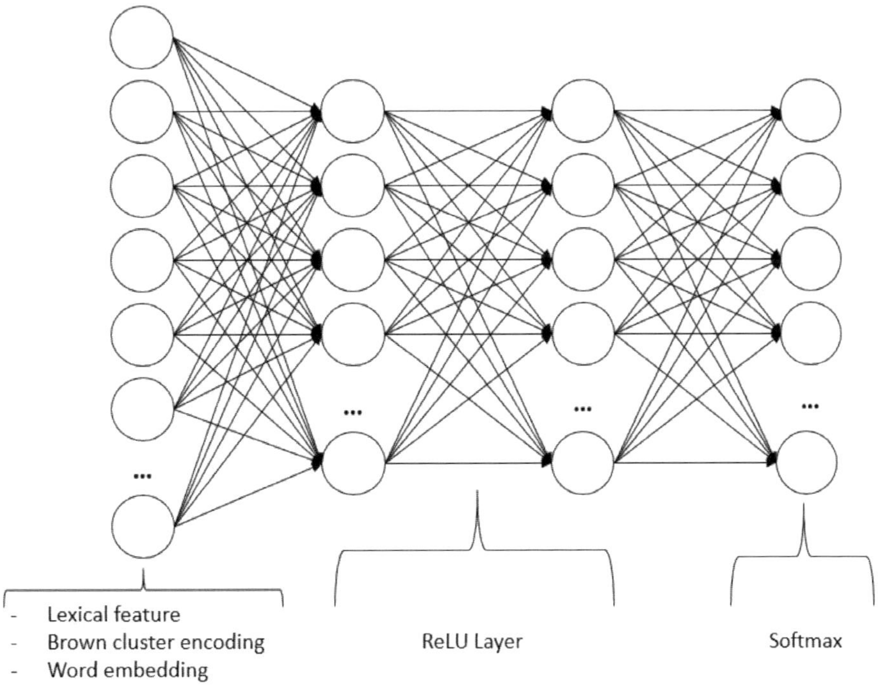

Figure 4. Proposed neural network architecture for the NER model.

We will train the network using the cross-entropy loss minimization as the training objective, with an addition of L2-regularization term. The neural network parameters correspond to the weight matrix W and bias b.

The input layer consists of baseline features, Brown cluster encoding (in the form of n binary input nodes), and/or word representation of the current word to be tagged. We use Zhang and Johnson's baseline features [15], which has been proven to be effective for NER tasks:

1. Previous two named entity predictions y_{i-1} and y_{i-2}.
2. Current word x_i, with several word type information: all-capitalized, is-capitalized, all-digits, and alphanumeric.
3. Prefixes and suffixes of x_i. The tokens between hyphens ('-') are used if at least one hyphen character is present in x_i.
4. Context $c = (x_{i-2}, x_{i-1}, x_i, x_{i+1}, x_{i+2})$.
5. Capitalization pattern in the context c.
6. Conjunction of c and y_{i-1}

We normalize dates and numbers when using lexical features (e.g., 2017 becomes *DDDD*, and 0752-618201 becomes *DDDD*-*DDDDDD*). This process is performed separately from the word representation usage. For example, if we have induced an embedding for 10/18/2017, we will use the embedding of 10/18/2017 and *DD*/*DD*/*DDDD* in the baseline features listed above.

4. Test Scenarios

We will test our work by comparing both the word representations themselves and the generated models. We will test the word representations performances using perplexity, which is a cross-entropy based measure borrowed from language modeling. In particular, we will test the performance of Brown cluster and several distributed word representations, which are shown in Table 1. Because of the word embedding's flexibility in its training parameters, we will also compare several word embeddings trained with different model types, window sizes, and embedding sizes.

Table 1. Test scenarios for word representations.

	Word Representation	Potential Parameter Changes
1	Brown cluster	-
2	CBOW-word embedding	Window size (2, 3, 4, and 5) Embedding size (50, 100, 200, and 300)
3	Skip-Gram embedding	Window size (2, 3, 4, and 5) Embedding size (50, 100, 200, and 300)

We will use the Brown cluster and the best CBOW and Skip-gram word embeddings from the previous experiment for the model training. We will compare several models which use different features. The baseline features are based on Zhang and Johnson's set of features [15]. Table 2 shows the feature combination used on each model. We will use the F1 measure as the main metric for our model performances. In order to mitigate the overfitting effect from potential lack of data for our model training, we will use the *k*-fold cross validation testing method, with *k* of 5.

Table 2. Test scenarios for NER model.

	Feature Combination
1	Baseline features
2	Brown cluster
3	CBOW-word embedding
4	Skip-Gram embedding
5	Brown cluster + CBOW word embedding
6	Brown cluster + Skip-gram word embedding
7	Baseline + Brown cluster + CBOW word embedding
8	Baseline + Brown cluster + Skip-gram word embedding

5. Conclusions

In this study, we proposed our theoretical NER models for identifying medical named entities in medical semi-formal texts. The word representations cited here can hopefully be used to enhance Indonesian medical named entity recognizer in semi-formal texts. We also hope that this research can be used to spur interests in natural language processing research in the larger scale.

Acknowledgments

This study is financially supported by the Institut Teknologi Bandung (ITB) under Program Penelitian, Pengabdian kepada Masyarakat, dan Inovasi (P3MI). We would like to acknowledge Kridanto Surendro (Information System Laboratory – SEEI ITB) for supporting the writing of this paper.

References

[1] Atrinawati L H and Surendro K 2009 Assessment for knowledge management readiness *Electrical Eng. and Informatics 2009 (ICEEI'09) Int. Conf. on* **2** pp 399–404.

[2] Friedman C 1997 Towards a comprehensive medical language processing system: methods and issues *Proc. of the AMIA Annual Fall Symp.* p 595.
[3] Aronson A R and Lang F M 2010 An overview of MetaMap: historical perspective and recent advances *J. Am. Med. Informatics Assoc.* **17(3)** pp 229–36.
[4] Savova G K *et al.* 2010 Mayo clinical Text Analysis and Knowledge Extraction System (cTAKES): architecture, component evaluation and applications *J. Am. Med. Informatics Assoc.* **17(5)** pp 507–13.
[5] Lafferty J, McCallum A, and Pereira F C N 2001 *Conditional random fields: Probabilistic models for segmenting and labeling sequence data.*
[6] Tsochantaridis I, Joachims T, Hofmann T, and Altun Y 2005 Large margin methods for structured and interdependent output variables *J. Mach. Learn. Res.* **6** pp 1453–84.
[7] Brown P F, Desouza P V, Mercer R L, Pietra V J D, and Lai J C 1992 Class-based n-gram models of natural language *Comput. Linguist.* **18(4)** pp 467–79.
[8] Mikolov T, Chen K, Corrado G, and Dean J 2013 Efficient estimation of word representations in vector space *arXiv* (*Preprint* arXiv1301.3781).
[9] Mikolov T, Sutskever I, Chen K, Corrado G, and Dean J 2013 *Distributed Representations of Words and Phrases and their Compositionality.*
[10] Aradea S I and Surendro K 2014 An overview of multi agent system approach in knowledge management model *Int. Conf. on Information Technol. Syst. and Innovation* (STEI ITB).
[11] Tang B, Cao H, Wang X, Chen Q, and Xu H 2014 Evaluating word representation features in biomedical named entity recognition tasks *Biomed Res. Int.* **2014**.
[12] Turian J, Ratinov L, Bengio Y, and Roth D 2009 A preliminary evaluation of word representations for named-entity recognition *NIPS Workshop on Grammar Induction, Representation of Language and Language Learning* pp 1–8.
[13] Bengio Y 2008 Neural net language models *Scholarpedia.*
[14] Bengio Y, Ducharme R, Vincent P, and Jauvin C 2003 A neural probabilistic language model *JMLR J. Mach. Learn.* pp. 1137–55.
[15] Zhang T and Johnson D 2003 A robust risk minimization based named entity recognition system *Proc. seventh Conf. Nat. Lang. Learn. HLT-NAACL 2003* **4** pp 204–7.